The
Medjool Date

A Rare and Exotic
Gourmet Delicacy

The AMAZING Story
of
The Fabulous

Plus 100 Medjool Date Recipes

Marc E. Paulsen

The Amazing Story of The Fabulous Medjool Date With 100 Recipes

COPYRIGHT © 2008 Marc Paulsen Press

Inquiries: 503-330-1226
or
F. Jaime, 1330-A Perez Road, Winterhaven, California 92283

⬆

REVISED EDITION 2008

First Printing

⬆

Text, photography and design by author.

ISBN 978-0-977473737-2-4

Cover photo:
Jaime Date Garden
Bard, California

About the author

Marc Paulsen's background began with building construction and expanded to include fine art dealing, interior design, airplane sales and real estate development. Mr. Paulsen has served on management assignments ranging from trade show exhibitions of national scope and museum consultant to project manager for high-rise office construction and large sawmill machinery installations. He has traveled numerous countries and, while serving with the U.S. Armed Forces in Germany, wrote many articles on tourism.

Marc's interest in the famed Medjool date came about as a "fluke." While enjoying desert sunshine in southern Arizona, he stopped to take photos of men working high in date palm treetops. A palm grove worker invited him to come into the grove to film the palms close-up. This book is the result of that simple original effort to take an unusual and interesting photograph. At that invitation, he began to learn of the Medjool's unique and surprising history and of the astounding amount of labor required to produce a crop of this legendary date. This book, then, is to let the world know this date's extraordinary story of survival against great odds and the details of how much human persistence and effort go into producing this delicious exotic gem of fruit.

Other books by Marc Paulsen:

Dominican Republic Guidebook
1989 - in advance of 500th anniversary (1492-1992)
of Columbus' discovery of the New World

❖

Flying Stories
From The Ragged Edge

❖

Audacious Escapades
In The Fabulous
Columbia River Gorge

❖

Short Stories
for
Quick Readers

Acknowledgment

A large measure of gratitude is owed my good friend Fred Jaime for his generosity and untiring patience explaining the numerous details involved in the production of a successful Medjool date crop. I also wish to thank Bard, California, date grower Ignacio Jaime for his kind hospitality and assistance; David Jaime for suggestions and his help placing me high up in the trees for photography; Alfredo Barajas for his helpful assistance up in the swaying palms and Greg Raumin, Jewel Date Company, for enlightening me with information about Medjool cold storage, Medjool retailers and date culture details in and around the Palm Springs/Indio/Thermal, California, areas.

Grateful thanks also to Glen Vandervoort of Vandervoort Date Ranches, Inc. who advised me on correct growing details and figures; Dave Nelson of Datepac LLC for connecting me with people able to shed light on Medjool history; Ronald Hill, Royal Medjool Date Gardens, for making available extremely valuable historic papers detailing important elements of the Medjool's incredible rescue story; Mari Nunez-Valleau, Imperial Date Gardens, Inc., for her helpful information and encouragement; Stephen McCollum, Sun Garden Growers, for his revision suggestions; Debra Mansheim, Bard Date Company, for her helpful insights. I wish also to thank the other kind people in the date industry who took time from their busy schedules to give me interviews so necessary to the completion of this book.

Lastly, my editor/sister, Diane Paulsen Heberling, deserves substantial credit for her untiring efforts in reviewing numerous drafts and providing much-appreciated journalistic expertise.

Medjool date palm workers spend many
hours high up in the tree tops

Society values uniqueness and high quality

It is a long-established fact that items of fabulous beauty, exceptional high quality and uniqueness of originality are among a society's most highly-prized and valued goods.

The Medjool date rates at the top in desirability

The "Medjool" date is a charter member of this rarefied fraternity of high value by virtue of its amazing history of propagation from antiquity, its decidedly matchless qualities as a food product and the intense difficulty of its cultivation and processing. It is not simply a "date" but rather the "Royal Superior" *of all dates!*

Growing Medjools is incredibly labor intensive

Medjool date palm trees require ten to fifteen years to mature to maximum production. Thoughtful, careful and precise nurturing is demanded at every step along the way. Few are aware of the astonishing degree of human labor necessary to realize a successful crop of this very unusual and delectable fruit. Dedicated workers ascend into the treetops numerous times during each growing season to perform these difficult tasks: pruning, de-thorning, pollinating, fruit cluster thinning, fruit arm training, individual fruit bunch bagging to protect from birds, insects and weather and, finally, harvesting. The thinning process alone is exceedingly time consuming. Working at top speed, one man can thin only six palm trees a day. There are numerous other necessary jobs below the spreading palms such as irrigating, field leveling, tilling, fertilizing, equipment manufacture and repair and even the trapping of root-eating rodents.

The famed Medjool "stands alone"

Over the centuries, the "Medjool " date was the date of preference for much of Arab desert royalty as well as others in positions of high station.

Dame fortune smiled on us

The Medjool's very existence in the California Imperial and Coachella valleys and in the lower Colorado River Basin of Arizona is the result of a surprising stroke of good fortune. It is possible that, in the entire history of world horticulture, there has never been a more amazing success story than the rescue of the great Medjool date from the sheer precipice of extinction!

Contents

Section I - Date Palm Early History

Section II - Medjool Propagation
(The comprehensive "technical" elements)

Section III - 100 Medjool Recipes

Section IV - Medjool Growers/Retailers

Jewel Date Company, Palm Desert, California

Section I

Date Palm Early History

The oldest tree crop

The date is considered to be the oldest known tree crop to be formally "cultivated." Dates have been an important food of the world for more than 6,000 years of recorded history. There are many references in the Bible regarding the importance of the date to early societies. Originally grown only in the Middle East, the date palm is a beautiful and productive fruit tree that provides the advantages of high food value dates suitable for simple long-term storage. As well as being the primary date-growing region in the world, the Middle East is also the largest consumer of dates of the more ordinary varieties. Of many varieties that have been imported to the United States, the regal Medjool is the most exotic of all.

Dates are self-preserving with long life

The basic properties and characteristics of dates provide a uniqueness that distinguishes them from all other major fruits. A mature tree-ripened date is self-preserving for months and can be stored as a concentrated source of food energy. Due to their relatively small size, they easily can be transported in containers small or large. As well as a great dessert food, dates constitute an important basic staple for many peoples of the world. In numerous harsh soil conditions, date palms flourish where other crops would prove marginal and, as a result, growers feel a special affection for the stately date palm.

As many as 400 to 600 varieties...and possibly more

The date palm was an early arrival in the western hemisphere. Most likely, it was introduced by explorers who came from tropical zones where palms were common. Some of their originally-planted palm trees or "offshoot" trees from those can still be found in Southern California and Mexico. Of the many date varieties that exist (some say as many as 400 to 600 and possibly more), the most unique of all is the legendary "MEDJOOL." However, the existence of the Medjool date in the U.S. is the result of an oddity of nature and a quirk of fate rather than a carefully-planned agricultural program of propagation.

ᛒow the Date Palm arrived in the Ú.S. Southwest

One man deserves much of the honor

Date culture in the United States southwest owes much to Frederick Oliver Popenoe. Frederick was born in Towanda, Illinois, in 1863. He first became a stenographer, then a court reporter and later worked as secretary to a governor. During his first forty years he mined gold in Costa Rica and worked as a representative for Pacific Monthly Magazine in California. In 1907 he moved to Altadena and opened a tropical plant nursery he named "West India Gardens" and introduced a number of sub-tropical plants and fruits to California. Some say his greatest achievement was the introduction of avocados from Mexico but those in the date business feel differently. Much of U.S. date culture is a direct result of major contributions made by Frederick and his two sons.

"F.O.'s" sons send palms from the Middle East

There is some question as to precisely what plants Popenoe's two sons, Paul and Wilson, were looking for when they traveled to the Middle East and North Africa between 1911 and 1913 but there is little question as to the significant long-term results of their trip. Biographers state that he sent his two sons there to search for tropical plants and date palms that might further his ambitions with his California nursery. Whether or not they found the specific plants they were searching for isn't clear but it is known that the brothers sent back 16,000 date offshoots from palms in Iraq, eastern Arabia and Algeria. These events effectively provided the impetus that resulted in the creation of the date industry in the U.S. Southwest of which "F.O." became a leader in California. His early efforts of successfully raising date palms paved the way for the later introduction of the illustrious "Medjool" palm.

Shields Date Garden about 1949
(Photo from a Shields Postcard)

Shields Date Garden, early pioneer

Shields Date Garden was established in the Coachella Valley near Indio, California, in 1924. It was one of the earliest agricultural operations to foster date culture in the American southwest. Their extensive palm grove operations, packing plants and retail and gift package store south of Indio rate among the largest distributors of the unique Medjool date. Shields is now part of a large business organization that has many date palm groves and raises a variety of date species.

A young Medjool date garden

A majestic mature Medjool date palm garden

Strikingly beautiful, the Medjool canopy spreads 25 feet

The Medjool Palm is rescued from the brink of extinction by visionaries and arrives in the U.S.A.

Dr. Walter Tennyson Swingle, 1871-1952
Botanist, U.S. Dept. of Agriculture

"To be told when he was born, at Canaan, Pennsyvania, in 1871, and when he died, January, 1952, is of less importance than to know he inspired more agricultural botanists than any other man." – William Seifritz

Morocco calls on the United States Department of Agriculture for help to save the Medjool

The Medjool palm might not exist at all if it were not for the vision and perseverance of outstanding Americans who stand exceptionally tall among their historical peers. They were the key players in an unparalleled achievement that is considered to be one of the greatest long-term success stories in the history of world horticulture...and we in the United States are among the greatest beneficiaries of that momentous event. Dr. Swingle was one of four people chiefly responsible for rescuing the Medjool date from extinction. Dr. Swingle acquired eleven disease-free Medjool date offshoots in Morocco in 1927 (when almost all Medjools there were dying) and brought them to the United States for an heroic and successful effort to save the Medjool from extinction.

Dr. Walter T. Swingle: To a date growers' association meeting:

"Early in May, 1927, by invitation of the French government, I joined a commission appointed to investigate the much-feared Baioudh disease of the date palm in Morocco."

Author: At this time the Baioudh disease was rapidly decimating almost every Medjool palm in that country. The Medjool palm was considered a "Date of Royalty" in Morocco. Most always had been reserved for royalty and those of high station. There were very few trees left and the species was considered on the verge of extinction when Dr. Walter Swingle arrived.

Swingle: "We entered one date garden after another only to find the Baioudh disease in or near every one of them. Finally we came to one that did not show any of the pale leaves in the middle of the leafy top, characteristic of the Baioudh disease. The gardens adjoining on three sides also showed no signs of it."

Author: Walter Swingle was a horticultural visionary. As a "Collaborator" of the Bureau of Plant Industry, U.S. Department of Agriculture, it was he who took the extraordinary action to save the revered Medjool palm from extinction. While in Morocco on this assignment, he acquired eleven disease-free Medjool palm "offshoots" for transplanting and sent them to the U.S.

The next heroic figure to enter the great saga of the Medjool's brush with extinction: Frank A. Thackery

Frank A. Thackery: (From a speech to an agricultural forum)

"In a Report of the Annual Date Growers' institute for April 28, 1945, Dr. Walter T. Swingle gave an excellent account of his importation of eleven Medjool offshoots from southern Morocco in Africa. They arrived at Washington D.C. on June 23, 1927, where the U.S. Division of Plant Quarantine gave them a severe vacuum fumigation. Following this, they were shipped to an isolated point in Nevada. On July 3, 1927, I received instructions to see that they were immediately planted in a thoroughly isolated location in southern Nevada. I was to select a place where they could be planted and cared for safely away from any other date palm. The extreme southern point of Nevada touches the Colorado river for a short distance about 22 miles north and up the river from Needles, California. Having never been in that portion of Nevada, I was completely in the dark as to what I might find there.

ᴀ second important figure joins the effort to save the ᴍedjool palm

Frank A. Thackery, longtime Indian Agent, played an important role as advisor to USDA involving Indian agriculture and joined the effort to save the legendary Medjool Date Palm as reported by him and printed here.

Early USDA field symposium on primitive agriculture, 1922

(Thackery is seated just left of the tentpole in center)

    I left the U.S. Date Garden alone in an old Ford Model T coupe on July 3, 1927, with the expectation of spending the fourth of July alone on the hot, dry desert of southern Nevada. The temperature was then ranging around 117 in the Needles area. At Needles I was finally able to contact by telephone a railroad man who gave me the location of the Medjool offshoots and I soon arranged for their immediate delivery in extreme southern Nevada near the Colorado river. I had expected to find it necessary to arrange with some Indian of the Needles locality to establish a camp in extreme southern Nevada and use his team, wagon and barrels to haul water from the Colorado river for irrigating these offshoots.

On the morning of JULY 4, 1927, I left Needles and followed a very dim wagon trail 22 miles up the river into this southern point of Nevada. I was impressed with the isolation of the locality. Soon after crossing the state line from California into Nevada, I was most fortunate in finding the hut of an old and badly crippled Indian named Johnson. His wife was also very old and nearly blind. There happened to be two younger Indian men there for a short visit. All of them belonged to the Chemehuevi tribe of desert Indians. One of the younger men, Mike Tobin, could converse fairly well in English. Through Mike, as interpreter, I made known the purpose of my visit. It soon developed that Mike had worked for my brother at old Fort Mojave and that they had enjoyed an extended and very friendly acquaintance. This helped me a lot in my introduction to these strange Indians.

Near Johnson's hut was an old dug well some twenty-five feet in depth— deep enough to tap the underground water from the nearby Colorado river. Over the well was an old broken-down wind-mill and pump. The well had caved in until it showed no water, but Mike assured me that a little digging would provide all the water required for the irrigation of the eleven offshoots.

Through Mike, as interpreter, I was soon given the approval of Johnson to put the well, pump and windmill in order and to use them for an indefinite time to water the dates. Johnson agreed to care for and turn on the wind-mill as needed to keep the offshoots well watered. He further permitted me to use about one-half acre of fairly level land situated slightly below the well where the water would flow by gravity to the offshoots. Nearby was an old fence with enough posts and wire to fence in the dates and thus protect them from the Indian ponies. Things were certainly coming my way. Here was a small patch of good, level soil, a well, pump, wind-mill, posts and wire plus an old Indian residing there permanently who was willing, even anxious, to care for the dates. Also there were two young Indians willing to help me with a very strenuous 4th of July "Celebration." Perhaps the patriotic 4th of July start given these dry offshoots may have improved their chance of survival, or perhaps the earnest prayers of that old Indian Medicine Man, Johnson, may have entered into it. Anyway, it has always seemed to me a miracle that all of the eleven imported and thoroughly dried out offshoots not only survived, but actually grew vigorously right from the start.

I was asked later just why I employed such a badly crippled old Indian to care for these important offshoots. There were several good reasons. First, because no one else was easily available. Second, because Johnson's old age and badly crippled condition assured me that he could not leave the place on frequent and extended absences to attend all sorts of Indian

A Medjool date palm "offshoot"

gatherings which have irresistible attraction for practically all Indians. Perhaps this may represent one of the very rare cases where old age and rheumatism are an asset rather than a liability. Another point in favor of Johnson was that old Indians of his type can usually be depended upon faithfully to keep an agreement. Also, both he and his wife were seriously in need of the modest compensation I was able to arrange for them. They attended to this job faithfully for approximately seven years.

We finally finished the cleaning of the well and the repair of the pump and wind-mill and had them working satisfactorily. It took us until nearly midnight to complete the planting, irrigation, and fencing. As I unpacked the offshoots after a long, difficult day in the hot July sun, I was discouraged and depressed because the shoots all appeared completely dried out and without life. I noted that seven large shoots showed several small secondary shoots— more than normal for large offshoots of other varieties. I also noted that both the large and the small shoots, including those still attached to the seven large ones, showed plainly the beginning of root formation, and that fact gave me slight hope for their survival. The four small shoots were about the size of one's fist and forearm. Seven years later when the entire planting was moved to the U.S. Date Garden at Indio I again noted that tendency to very early formation of roots, even on very small shoots having no soil contact. I made my

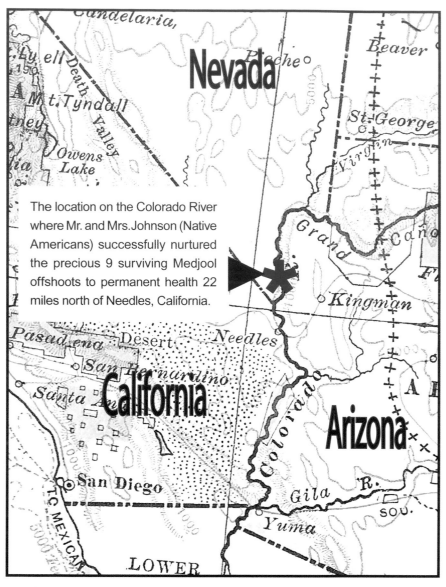

The location on the Colorado River where Mr. and Mrs. Johnson (Native Americans) successfully nurtured the precious 9 surviving Medjool offshoots to permanent health 22 miles north of Needles, California.

The location on the Colorado River at which Native Americans nurtured eleven endangered Medjool date palm "offshoots" for *seven years* (only nine survived) and therefore played an extremely critical role in the Medjool date palm's rescue from extinction.

first visit back to this isolated planting approximately sixty days later. I was astonished to find definite signs of life and even slight growth on each of the eleven imported offshoots. I encouraged Johnson in his excellent care of the planting and it pleased him very much when I told him that he must be a real "Medicine Man" as indicated by his ability to make sick plants grow. He told me that he had asked the "Great Spirit" to help him and I noticed that he increased his determination to do his very best. A few months later, however, he became ill and could not get out even to turn the wind-mill on and off. His wife, who, as stated, was almost blind, cared for the irrigation during Johnson's illness. Because of her very poor eyesight she had not noticed that, during a few extremely hot days, their dogs had dug up two of the smaller offshoots in trying to find a cool, moist bed. Some weeks later when Johnson was able to resume his work, he discovered what the dogs had done. He thought these two shoots had been exposed to the hot sun until they were completely dried out and were dead so he burned them on his camp fire. Thus the number of imported offshoots was reduced from eleven to nine.

These nine remaining shoots were kept in isolation for more than seven years. During that time, they were inspected several times by State and Federal officials including Dr. H.S. Fawcett, Dr. Walter T. Swingle, Mr. A.J. Shamblin and myself. No disease or insect pests were found on them at any time. Finally, it was determined that it would be safe to transplant them in proximity to other date palms at the U.S. Date Garden. Mr. George H. Leach was assigned this important task. With the idea of increasing the chance of the originals to survive, he first removed sixty-four offshoots, many of them quite small. Most of these sixty-four offshoots were taken from the seven larger imported shoots. These sixty-four new offshoots plus the nine imported offshoots (by this time young palms) made a total of seventy-three to be transplanted. The fact that every one of these survived not only was evidence of expert work by George Leach, but was additional evidence of the unusual hardiness of this date variety. Added to this is the fact that all of the eleven imported offshoots survived the long trip to the United States from Morocco and a severe vacuum fumigation on their arrival at Washington.

The young palms were moved to the U.S. Date Garden at Indio, California, in the summer of 1935.

Author's note: This amazing account of the Medjool Palm's "hair-breadth" rescue from absolute disappearance on the planet is one of the most exciting stories and displays of vision and persistence ever to take place in the entire world of horticulture. If this great story is not dramatic enough, consider this additional situation Thackery also resolved: After working out the offshoot care and irrigating details with Johnson, the old Indian, it was discovered Johnson's property was just off the Indian reservation...and he had no title to the land! Thackery somehow managed to get the Reservation boundary changed to include Johnson's property.

Now....that's determination!

A recently planted offshoot with sunburn wrapping. This is how Native American Johnson protected the eleven offshoots from the harsh desert climate.

Young palm tree with new offshoots

Most Medjool palms in the world are from the nine surviors of the eleven original offshoots brought to the U.S. to be saved from extinction.

Those surviving nine of the original eleven offshoots are the origin of all Medjool palms in the United States. In fact, in later years when Medjools had recovered sufficiently in the U.S., young "offshoots" from the new healthy trees were sent back to the Middle East and now most, if not all Medjools grown successfully in the world, are the offspring of those plants saved in that victorious rescue effort. Plantings from the offspring of the original offshoots were later made in Texas, Florida, Arizona and a few other locations to determine just where in the U.S. the trees would be most successful. Ultimately, it was learned through trial and error that dates grow best in areas having exceptionally high summer temperatures, very low humidity and abundant water for their root supply. Experimentation proved that only a very small southwestern section in the U.S. provided the best all-around growing environment for the great Medjool date.

Medjool production is very limited in the United States

The Medjool date-growing areas in the U.S. are the Bard, California, area of the Imperial Valley, Arizona's Yuma County and the Palm Springs-Indio-Thermal area of the Coachella Valley of southern California. Production of Medjool dates is as limited in the Middle East. A fact that surprises many is that a portion of the U.S. Medjool crop is exported to several Middle Eastern countries. Another limiting factor is that Medjool production requires growers to have a sizable investment in cold storage and freezer capacity. The Medjool is much softer (the characteristic that makes it so desirable) than other varieties and for that reason requires much different handling including more complicated refrigeration. The Medjool freezes well and that is just one more benefit for the consumer.

Last of Dillman's 1944 planting
Stanley Dillman: A Medjool palm culture pioneer

In 1944, twenty-four offshoots, all descendants from the original nine survivors, were planted in the Bard Valley of California near Yuma, Arizona, by date pioneer Stanley Dillman. Dillman carefully nurtured his palms, experimented with various growing and harvesting methods over a number of years and ultimately became an early self-taught highly knowledgeable U.S. Medjool palm "field" expert. He wrote a number of informative papers describing his methods...and shared them generously. Many of his ground-breaking Medjool date culture innovations are still in use today.

Some Medjools grow surprisingly large

The MEDJOOL - *very different*...and...*very large*

By those closely familiar with date culture, the Medjool has long been considered the unquestionable "Royalty" of the entire date family and is certainly one of the most important date varieties now grown in the United States. The texture, flavor and appearance, plus its great size (about an inch by two inches...twice the size of ordinary dates) make the Medjool by far the most impressive of the date family. It is dark reddish-brown in color with a skin appearing somewhat wrinkled. The dark amber flesh is rich, soft and chewy and seems to "melt" in one's mouth. Adding to nature's special attributes given the Medjool, compared with many lesser counterparts, its success is due in large measure to the wonders of modern horticultural practices and the considerable amount of extraordinary effort put into its production.

To begin, climate is almost EVERYTHING!
Date palms require very hot weather

Medjool date palms require an exceptionally hot climate. That is precisely why most dates are grown in the hottest desert climates of the world. The U.S. growing areas of the Coachella Valley, the Imperial Valley and the Lower Colorado River Basin have climates that are very similar to those of the other major world date-growing regions. These areas have much of the finest and most pleasant moderate winter climates in all of North America. However, the middle of summer in these areas is a very hot time indeed.

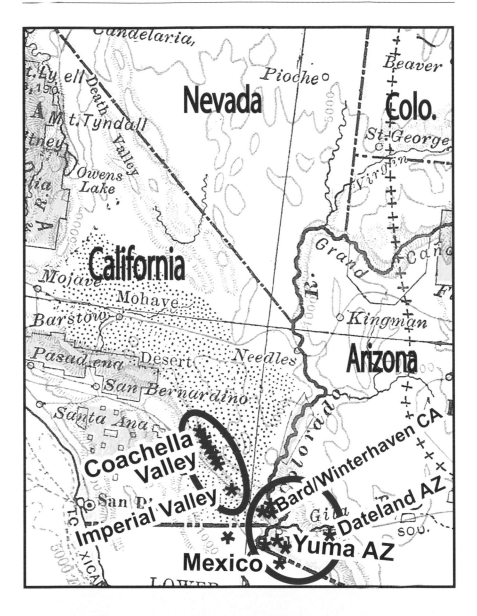

* The major Medjool Date growing
areas in the United States and Mexico

The Medjool Date is a very large, luscious, soft and flavorful candy-like natural confection

U.S. growing areas are hottest in the country

The key fruit-ripening period for Medjools in both the lower Colorado river basin and the Coachella Valley is from June through September. During these months the temperature averages over 100 degrees on a regular basis. July and August are the hottest usually averaging over 105 degrees per day for the entire month. When harvest begins about the first of September, the workers (called Palmeros) must spend their days in the trees working in temperatures so high that most people are scrambling to find air-conditioned places to make the weather bearable. But the harvest must go on and the Palmeros are the people we owe our thanks to for facing these severe conditions and making this fabulous fruit available for our epicurean delight.

Medjools are a very complex crop to grow

The complete method of growing Medjool dates is little understood by casual observers. Few would ever imagine the extensive and complicated amount of human and mechanical effort required to produce a successful crop of this wonderful fruit. If they had solid knowledge of the tremendous expense and human physical effort involved, they would realize the price asked for this product by their growers is very reasonable indeed.

Oasis Date Gardens, Indio, California

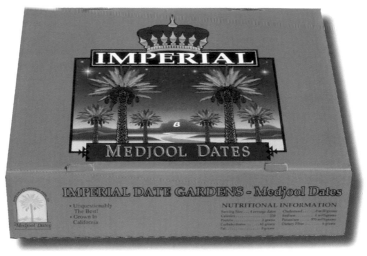

Imperial Date Gardens, Bard, California

Placing Medjools for air "firming"
One of many processing activities

Section II
Medjool Propagation
(The comprehensive technical elements)

Tasks necessary to grow a Medjool date

- Dethorning branches
- Collection and disposal of thorns
- Harvesting pollen from male trees
- Hand/machine-pollinating female trees
- De-centering new fruit arms
- Thinning/tying/training fruit arms
- Pruning of "offshoots"
- Removing offshoots (when applicable)
- Pollinating late blooms
- Removing late thorns
- Irrigating (numerous times)
- Fertilizing (organic)
- Tilling (keeping land neat, free of weeds)
- Thinning fruit strings/training fruit arms
- Removing old fronds
- Converting old fronds to mulch
- Inserting spreader rings inside clusters
- Bagging fruit arms
- Harvesting (3 to 4 times up into the trees)
- Removing bags
- Packinghouse activities
- Removing spent fruit arms after harvest
- Converting fruit arms to mulch
- Washing/repairing/making bags
- Preparing/renovating required equipment

Dateland Village, Arizona

The Palmero's view

℘almeros

The workers who make it all possible

Just a "Day at the office"

A very unusual type of worker

To work in the treetops, Palmeros are raised up on platforms by either high-lift multi-stage fork trucks or hydraulic-extension-boom vehicles. Some tasks can be done while simply walking on the platform when raised high up into the tree fronds; however, a large share of the work they do involves climbing from the raised platform on up into the treetop then continuing their efforts by standing directly on the tree branches for support while stepping and working around the treetop.

Men of little fear

It takes a special type of person to be a Palmero. It is an amazing sight to observe them from the ground as they ply their craft high in the palms. Although required to tether themselves with safety lines, there is always danger when changing locations of the safety gear or walking on wet branches or working in high winds. There is also a danger of falling when transitioning from a lift vehicle to the palm treetop. For those reasons, a worker must have considerable self-confidence and be a thoughtful and carefully calculating person. There are few second chances for those who make major miscalculations in their movements on high. The people who do the actual manual labor high in the trees are a valuable element of the entire date-growing process.

Palmeros perform a myriad of different tasks

It takes 10 years of devoted care before date palms reach their full production of about 100 to 200 pounds per tree per year. Mature treetops grow from thirty to fifty feet in the air. Tasks that take place at different times of the year are de-thorning frond stem branches, pollen harvesting, hand-pollinating, cutting away dead fronds, de-centering and fruit-string reduction of the fruit arms, tying down fruit arms to facilitate ease of harvest, insertion of round wire spreaders to hold fruit apart and allow greater exposure to the open air and warmth of the sun, thinning of individual fruit strands, attaching bags over the fruit clusters to protect them from weather, birds and insects, and, finally, three or four separate very selective pickings of the ripened fruit. The actual harvesting of the ripe fruit is much more complicated and labor consuming than a casual observer might think. Each palm must be climbed as many as 18 times a year to carry out all these manual operations. To achieve optimum size and high quality of the tender ripened Medjool date, special care must be exercised in every step along the very lengthy path to final perfection. These unique Palmeros must be precise about each and every task they perform.

Training is all-important

It is necessary to have highly-skilled Palmeros with considerable experience impart their knowledge to trainees. The quality and yield of a grove depends upon workers completing tasks in exacting fashion. The proficient trainer must demonstrate much patience in the worker-training process. The jobs of fruit arm de-centering, fruit string thinning, fruit arm down-training and the thinning of the actual fruit on fruit strings require critical attention to detail. It must all be done right *the first time.* There are no *simple* manual tasks. Experience can come only from careful on-the-job training and then long exposure to the work by those who are 100% fearless of height and at home in a treetop among unstable branches...and who are alert every moment as well as meticulous to a fault.

Palmero up in tree pollinates a Medjool palm

Only a select few "make the grade"

Young Palmero applicants are carefully assessed to ascertain their level of physical fitness as well as their natural agility. A strong sense of balance is an absolute basic requirement to perform work in the palm tops. They also must be able to concentrate intently on the job at hand and, at the same time, be prepared to act instantly (and more or less automatically) to correct any situation that might indicate approaching peril. For example…if they accidently drop a tool or item of equipment, they must know when to reach out quickly to retrieve it or just let it go and maintain their equilibrium in the tree. If they can demonstrate unusually good balance, then they are quizzed to determine just how they feel about operations at great heights above the ground. They must be completely comfortable with heights to be effective in this line of work. Palmeros deserve a tremendous amount of respect. Without these fearless toilers in the sky, no one would be able to relish the candy-like Medjool date.

Palmeros are fearless *"Skywalkers"*

Medjool Propagation

A young Medjool grove

Date Palms are as old as history

From ancient times, the Date Palm (*Phoenix dactylifera*) has been culti vated in the deserts of North Africa and the Middle East as a chief source of food. Information excavated from 5,000-year-old sites tells us the date was a highly-prized staple of the ancients. Fossil records indicate the date palm is millions of years old.

More than 600 varieties of "common" dates

According to horticulturists and data keepers, there are as many as 600 and more different varieties of dates and the world's annual date harvest exceeds a few million tons. The Food and Agriculture Organization of the United Nations has estimated that, of 90 million date palms in the world, 64 million grow in Arab countries mostly in the desert regions of North Africa, the Arabian peninsula and southern Iraq where dates have been a staple food for millions of people over thousands of years. This is because they perform best between 15 and 35 degrees north latitude which encompasses the world's hottest, driest deserts. They are relative newcomers in the southwestern deserts of the United States.

Conquistadors did more than plunder

By most accounts, Spanish Conquistadors were first to introduce the date palm to the New World. Later, during the settlement period, Spanish missionaries were known to establish date palm groves in the areas where they built the first permanent missions. Those dates were of the ordinary species. Most of the early Spanish missions where they were first planted were located relatively near the coast and this higher humidity location proved to be generally unfavorable for the most successful date cultivation. Over time, growers in those areas gave up and all date propagation moved inland to desert soils.

Date palms like it hot

Date palms cannot survive in shade and must grow in areas of almost total sunlight. They will grow in all warm climates where the temperature averages well above sixty-seven degrees to allow the males to flower and females to fruit. The date palm can withstand lower temperatures during dormancy but not for long periods. They grow best in extremely high summer heat with little rain. They can survive long periods of drought but, to achieve high production, their water requirements are very high. The date palm thrives best in light sand and sandy loam and requires good drainage and aeration.

Too much salt can stunt growth

Date palms can tolerate a surprising degree of alkali. Some salinity can be endured but too much salt will ultimately stunt growth and lower the fruit quality. Excess salinity in the lower Colorado River water has been a major problem for many years because the construction of the giant dams on the river has altered the chemistry of water that flows downstream from them. This has adversely affected many farm-

Offshoots crowd around a parent tree

ing operations that depend on the massive amounts of irrigation water diverted from the Colorado River. This continuing problem is addressed with drainage systems and other management programs that attempt to minimize the salt's effects on those agriculture areas which include *most* of the Medjool date grove operations in the U.S.

Date palms are male and female

Date palms are "dioecious", each tree being specifically male or female. A date palm can grow from a seed but the resulting fruit is usually inferior to that of the original palm in size, flavor and taste. All date seeds typically produce a completely new variety of date that most likely will be substantially different from the fruit of its parent tree. At the base of each palm leaf is a single bud (ancillary bud) which is formed at the same time as the leaf. In relatively young palms, some of these buds may differentiate vegetatively into offshoots when the leaves are about three-quarters of an inch long. Several years elapse before the offshoot grows sufficiently to emerge from the surrounding leaf fiber.

New trees come from offshoots

To ensure high quality dates, it is crucial to obtain new trees in the form of offshoots from carefully-selected mature female palms of proven fruit-bearing quality. The parent date trees, both male and female, produce "children" in the form of offshoots that spring from their base and will always be identical to their ancestors.

Therefore, to perpetuate the identical species, it is necessary to separate offshoots (which have the same genes) from the parent tree and transplant them to create a new grove. When an offshoot grows at the base of the parent tree to begin a second generation, it can be removed after three to five years of growth when it has developed a substantial root system of its own at the side of the parent tree. When separated from the parent, offshoots will weigh from 35 to 100 pounds usually averaging near fifty. It will be ten or twelve inches in diameter and have a few branches with a height of four to six feet when trimmed for transport. An additional benefit of removing surplus offshoots is that maximum food value is then allowed to flow to the parent trees. Offshoots are removed and transplanted in the late spring and early summer. In many situations they are transplanted to a holding area for a couple of years until they have achieved enough vigor to grow to maturity and are then moved to a final grove location. If a date happens to fall at a tree's base and the seed from it germinates, it causes a degree of confusion because it can spring up relatively unnoticed among the other offshoots. These must be spotted and removed so they are not mistakenly transplanted along with the true offshoot descendants of the tree.

A "baby" palm tree tied for removal

Offshoot separation is a very delicate process

The root system of the offshoots intermingle with that of the parent tree so, when they are separated, it must be done with great care to retain most of the offshoots' critical growth but not damage the parent's root system too much. The offshoot is connected to the parent tree out of sight below the surface at its bottom by what some Palmeros refer to as an "umbilical" connection. If offshoots are allowed to grow too large against the female tree, it is possible they might actually begin to push the parent tree over and out of a vertical position. To eliminate this situation, the offshoot will be removed even if the grower has no use for it. Offshoots can become so numerous they make it difficult for workers to perform needed tasks in the trees (mostly in younger trees that are not very tall but are producing fruit). They then remove those offshoots which have the least potential to make successful new trees. A constant yearly pruning regimen is required to allow easier access to younger fruit-bearing trees and to permit the most nutrient value going to the parent tree.

Offshoot roots are carefully exposed

The soil must be moist and soft

It is best to remove the offshoots when the earth around the trees is quite moist. This allows for easier digging and also helps hold soil to the removed offshoot root system for a more positive transplant outcome. The connection must be located and then cut as close to the parent tree as possible to cause the least damage to either plant. First, Palmeros carefully dig and remove the earth around the base of the offshoot down to a couple of feet or so below ground level trimming off its small superfluous new roots as they go. When the earth has been cleared all around the base of the offshoot and the "umbilical" located, a small rope is fastened to the offshoot's top and pressure applied to pull it gently away from the parent tree. Finally, the Palmeros apply a specially made four-inch wide long-shaft chisel to the umbilical and slice the offshoot free. When

Grower-made offshoot separation chisel

the offshoot has been removed, the large hole created for its removal is then backfilled. Even though a tree may have as many as six or eight offshoots, Palmeros are careful to remove only one or two at a time so the mother tree will not be weakened.

Growers make their own tools

Special tools are necessary for the removal of offshoots. Growers must make some of their own tools due to the unique nature of the offshoot removal process and the lack of factory-made tools for this limited work. Growers manufacture their own special date palm offshoot-removal chisel which is the key tool for a successful operation. The chisel blade is of hardened steel about six inches long by three to four inches wide welded to a steel shaft four or five feet long that is typically an inch in diameter. One side is flat and the other beveled much like an old-fashioned broadax. Offshoot removal usually requires two workers to be efficient. One holds the chisel in place and the other drives a heavy sledgehammer to force the separation.

Offshoots are trimmed and transplanted

In a special area, the offshoots' bottom growth structures are cleaned up by trimming the remaining loose roots and cutting free any lower unnecessary branches and growth. They are then immediately planted and watered in a temporary nursery area that has ready irrigation. The branch system above ground is completely wrapped in cloth to inhibit "sunburn" as the plants adjust to their new growing environment. The offshoots will be left here to gain strength and grow their own root system to a size that will facilitate their next move to a final growing place in a permanent grove. Usually, that period is a couple of years.

Medjool is a fruit of limited availability

Although date production is extremely labor intensive, worldwide production has increased substantially in the last 30 years largely due to improved date-culture knowledge and newly instituted labor-saving devices and equipment. However, it is still a fruit of very limited availability in the United States due to the relatively small acreage devoted to its production, highly-restricted

Driving the large "separation" chisel

locations for its successful propagation and a fairly small number of people willing to dedicate their efforts to such a demanding process. In addition, Medjools require a very special type of worker to toil so high so often and long in the exceedingly hot areas where they grow.

Date growing locations *must have* just the *right combination* of basic desert sand/sandy-loam soil conditions and an average growing season summer temperature that ranges well above 100 degreesFahrenheitwith minimal rainfall and low average humidity. It will be seven to

Separated and ready for trimming

ten years before a somewhat mature crop of dates will be harvested from these trees. Only farmers with the highest degree of patience and persistence will be successful. There is a very small amount of property in the contiguous United States that provides the basic necessary requirements of the Medjool date palm and most of that is already under the intense cultivation of other high-value competitive vegetable crops.

Each offshoot is
carefully trimmed

A final move

After a couple of years when an offshoot has established itself with its own larger root system in a "nursery," it is dug up and "balled and wrapped" tightly in burlap or cloth to hold the sand-silt earth in place around it. This allows transportation movement without endangerment by losing soil from the root system.

Offshoot is wrapped to
prevent sunburn

Planting 1st Medjool in a new plantation

Medjool palms are planted up to 30 feet apart

Young palms are planted with wide spacing which usually works out to 50 trees per acre. This varies somewhat from grove to grove but, in general, about thirty feet apart is the norm. Offshoots begin to bear modest amounts of fruit in three to five years and reach mature bearing age and full production nearing ten years. It is possible for a date palm to continue producing for as long as a hundred years although studies show that production declines after 50 to 60 years. The palm grows at a rate of about 1 to 1-1/2 feet per year and can reach 20 to 30 feet in 15 years depending on how it has been attended.

A new plantation takes form

Medjools are a beautiful decorative tree

Medjool date palms make an impressive statement when planted for pur-
poses of landscaping; however, harvesting dates from such a decorative tree is
a losing proposition because so much work is required to obtain a crop anything like that achieved by commercial growers. It just isn't worth the substantial effort demanded for one or a relatively small number of trees. Most
who have them for decoration simply let the birds have their way.

✒*Irrigation*

Date palms require much water

Even though date palms are a desert plant, they require substantial amounts of water. The root system is fibrous and somewhat similar to that of a corn plant. Roots about 5/16 of an inch in diameter develop from the outer part of the palm trunk and grow outward and downward. Small secondary roots develop from these main roots. While roots have been found as far as 78 feet away from palms and over twenty feet deep, about 85 percent occur in the upper 7 feet under normal development in a deep sandy loam soil. The date palm, much like the willow, requires water that a rainfall of 100 inches or more a year would provide. In the U.S., irrigation is the key to success. For that reason, date palms are found only in the U.S. southwest where artificial irrigation systems, both "flood" and "drip", are in place to make possible the large amounts of water required.

Plenty of sun is imperative

Date palms grow where two things must be in great abundance...very hot sun to transmit rays of energy to swaying fronds *plus* prodigious amounts of water available to their demanding root systems. Sages of the romantic deserts of yore would proclaim, "The date palm must have its feet in water and its head in the fires of heaven."

Much water comes from the great Colorado River

The date palm groves in California and Arizona are the happy recipients of water that was initially diverted mainly from the Colorado river for the purpose of helping settlers farm successfully in the early U.S. West. Irrigation of the West got its biggest initial boost in 1902 when congress passed the "National Reclamation Act" and authorized the Secretary of The Interior to develop irrigation projects with government money, the cost to be repaid by the recipient landowners. Although several private canal and pump companies had been started in lower California and Arizona as early as 1897, it was this act of the government that provided the impetus that set in motion monumental construction efforts to make large amounts of water available to farmers on a continuous and reliable basis. Without substantial amounts of "borrowed" water necessary for date culture that takes place in only these most arid parts of the contiguous United States, you would not be enjoying these greatest dates in the world.

Appropriate moisture content is critical

Too much rain at an inappropriate time can be devastating. If late summer rains are accompanied by too much ambient air humidity, damage to ripening fruit can be substantial…especially if the grove has just been irrigated, a situation that exacerbates the problem of moisture content in the air around the trees. Fruit quality is best when the trees have *moderate* moisture during ripening. If soil moisture content is not maintained at a reasonably high level, reduced size of fruit occurs. This is just one more "balancing act" for growers. Where relatively heavy irrigations are continued throughout the year, total water requirements may be six to eight "acre-feet" of water. Large dates require lavish watering!

Many groves are inundated by "flood" irrigation

Much irrigation is still done by the "flood" system. Water is distributed in "furrows" or by flooding the entire field. Furrows are constructed between rows and water let in on a highly-controlled basis. To flood, the palm groves are planed (leveled) as close as possible to plus-or-minus an inch or so of perfectly flat. The edges of the groves are banked just enough so the entire grove becomes a de facto "basin." This eliminates wasteful runoff. Water is turned into the grove and allowed to stand until it percolates down to the root systems below. No work can be done in the grove when flood irrigation water is standing. Because most operations involve high platform work, the earth must be very solid so when platforms are raised the vehicle used will be solidly stable. A slight drop of one wheel at ground level translates into a substantial side-movement of a work

A young grove flooded

platform high up in the trees. It is tricky to irrigate in this fashion and still calculate exactly when the water will have dissipated to the point that wheeled vehicles can navigate through the trees without either getting stuck or tipping precariously when one wheel might sink more than its opposite. During times of intense irrigation, growers perform a juggling act to balance applying the necessary large amounts of water and still preserve enough dry-ground-time to accomplish the mandatory tasks up in the treetops.

An alternate irrigation method is "drip"

In situations where "flood" water is not available or it may be more efficient to apply water directly to each tree, the "drip" system is applied. There are circumstances and conditions that can make drip more convenient or advantageous. Because most of the grove is dry between trees, drip irrigation operations make it possible for operation equipment to enter the fields at will or, at least, more frequently than flooding. Drip irrigation generally requires much expensive equipment in the form of wells, pumps, piping and a lot of attention to making sure the waterflow is continuous to every tree and not inhibited in any way.

Typical "drip" irrigation system
(note slightly visible lines running between trees)

Date palms are drought-resistant and can survive very long periods without irrigation. They respond very well with regular deep-watering especially in very sandy soils. As with other types of drought-resistant plants, they use water lavishly when it is available and the happy result is high production. To achieve and maintain maximum growth, the soil must be wetted to a depth of six feet to as deep as ten feet at least once during the winter and spring followed by regular summer irrigations at intervals of 20 to 25 days or less. It takes approximately eight inches of water to moisten the top six to ten feet or four to five inches of water to wet the top three or four feet. To gain the most growth in young palms, irrigation should be heaviest in winter months followed by frequent lighter summer irrigations.

ᴀ year's production begins

1st...The "de-thorning" process
Palm branch thorn removal...late
December through Mid-February

New thorns have tire-piercing potential

First treetop operation of the year

The first operation of the year in the treetops is to remove viciously-sharp long thorns that appear on the new branch growth. Although "de-thorning" usually begins in January, a few growers may begin this operation as early as late December.

Dangerous thorns must be removed

After being sliced off, the thorns fall to the ground below where workers on the ground rake them together <u>very carefully</u> into piles for disposal. This is necessary because the thorns are so tough and rigid they pose a danger

and easily can pierce human flesh or the rubber steel-belted tires of the assorted trucks, tractors and lifts that must work in the grove. Any thorns inadvertantly left on the ground remain strong, tough and lethal for a long time.

Another "job specific" tool

As with their specialized offshoot removal tools, the growers must also make their own de-thorning knives. The de-thorning knives are made from assorted hard steel items such as farm disc blades or rolling "coulters" from old implements. They are reformed into the curved shape seen above and heat-treated to hold a fine edge.

Grower-made "de-thorning" knife

Palmeros exercise extreme caution to avoid injury

High-lift platform at work

And...just one more "fringe" task

Catching and removing Pocket Gophers

Pocket gophers proliferate in some areas and have the potential to cause much damage to the root systems of palm trees resulting in direct financial loss. The little critters treat the roots of the Medjool palms as a veritable feast and do their best to build homes in burrows directly under the trees inside the root systems. Uncontrolled, they will expand their area of damage to where it will adversely affect the quantity of fruit production and they must be removed.

Palmero
sets trap

Rodent
peeks out

Got 'im!
One less root destroyer

Trapping while de-thorning

As Palmero crews work up in the trees, those with support jobs on the ground will sometimes do "fringe" work of gopher trapping. One successful method used is a simple "snare" arrangement using nylon fishing line.

Medjool date grower Fred Jaime displays very large pollen pod

Bard Valley Medjool Date Growers Association

ᛒarvesting pollen - ᚠebruary

Male pollen pods just before blooming

All pollination is grower-applied

Artificial pollination of date groves is necessary because the traditional method of letting the wind do the job is far too uncertain. Pollination is a critical two-step process. First, pollen is "harvested", then applied manually or by machine.

Male pollen flower ready for immediate "harvest"

Medjool palms are male and female

Only the female trees bear fruit. Male trees only produce pollen. The male trees grow pollen "pods" similar in appearance to the female but generally about one-third to one-half taller in height. The pods, brown in color, are about three plus inches in width near their center and range to five feet in total height. The fruit pods in the female trees and the pollen pods in the male trees look very much the same until they bloom. Then, the female trees produce a fruit "arm" flower made up of many fruit "strands" *or* "strings" that grow together in large bunches and the male trees blossom a "flower" of white pollen only. The males must be included for pollination of a grove.

Just 1 male tree to 49 females

When grown as a commercial crop, the typical ratio of male trees to female trees is one male to 48 or 49 females. In most grove configurations, male trees are planted on the perimeters to allow easier observation of their pollen "pod" flowers when opening and also ease of access for harvest. A typical mature male tree will produce about fifteen to twenty-five pollen pods and often more. Young male trees may generate only five to ten pollen pods.

Male pods bloom before the females

If normal weather conditions prevail, the male pods bloom about two to three weeks before the females. The male pods mature and bloom over the course of about a month and workers continually observe the trees as pods successively open to reveal their pollen flowers. Pollen harvesting from the male trees usually begins about early February. As the males bloom, the pollen pods holding their white pollen flowers are cut from the tree which means many visits to each male tree. When a pod pops open, a worker immediately scales the tree to cut that pod and flower out to minimize the possibility of pollen loss from wind movement or bees. The pollen is then processed into a dry powder. This process can last for many days.

Pollen easily can be lost

As part of the removal process, a large plastic bag is very carefully placed over the top of the pod and flower while it is still in the tree and tied at its base to reduce pollen loss through vibration caused by further movement. The flower with its stem and the flower's pod sheath are then cut free together at their base and removed to the location of pollen processing.

The pollen flowers are "processed"

The pollen flowers (still on their branch stems and in their pods) are then taken to a dry-room where they are removed from their pod sheath and held "flower-down" in a very large container such as a fifty-gallon drum or a large garbage can. They are then beaten against the inside of the container to shake the pollen free. The loose pollen that accumulates on the bottom of the container is collected and placed on a dry surface then left to dry for a day or two or until dry enough to sift through a screen to remove large particles. The drying is sometimes accelerated by applying heat. For tree application, the pollen must be absolutely dry and the particles very fine in size because it may be applied by hand-held shaker in some situations.

Pollen flowers are processed into powder

The processed flowers are saved and set aside

The pollen flowers (with almost all of the pollen now removed) are set aside to dry and are then cut into small pollen "twigs." In most situations, when the Palmeros apply the pollen to the fruit arms by shaker, they also insert two or three of the spent pollen twigs into the fruit arms to further guarantee pollination because the twigs still contain minute quantities of pollen residue that will reinforce the process.

Some excess pollen is stored for the next season

Under ordinary circumstances enough pollen is collected to allow for processing a surplus and storing it under refrigeration as a "holdover" supply for the next growing year. This acts as a form of pollination insurance. If the next year arrives with unusual weather that may cause male pollen pods to bloom too early or too late, then enough pollen will be on hand early to fertilize the females with less regard for matching the year's pollen harvest with the timing of the female flowers. The estimated mortality rate of "holdover" stored pollen is in the range of just a few percentage points so enough survives that the system usually works adequately. If there is a large surplus of stored pollen and it isn't all needed, it may be mixed with a current year's needs and fresh surplus pollen stored in its place for the next year. This assures that stored pollen is always no more than one year old.

Harvested male flowers ready to be broken
into twigs for insertion into female fruit arms

Far up at the top, a pollinator begins his day

Grower-assisted pollination

February and March

In days of old, the wind did the job...(hopefully)

Surprisingly, nature seems to have made no adequate provision for the positive pollination of the female palm tree fruit flowers. Originally, in the desert regions where the Medjool grew in a natural setting, pollinization of female trees was unreliable. With male trees somewhere nearby, miniscule particles of pollen would be transported about in the air during male tree flower-blossom time. Unfortunately, in some years there was not enough floating about to adequately pollinate all female trees in the area. This all changed when Medjool dates began to be treated as a domesticated man-cultivated crop. Then it became absolutely mandatory to be as sure as possible of complete pollination every year to guarantee a profitable crop (or maybe...any crop at all).

Nature needs a little help

"Grower-assisted" pollination of Medjool date groves is necessary because the traditional method of letting the wind do the job naturally is far too uncertain… especially when very large grower investments of labor and money are at stake. The most positive assurance of producing fruit every year is by helping nature along.

These days, growers give nature a "boost"

To achieve optimum date production in a "farm" setting, all pollinating now is done manually or by men and machine. Each individual fruit arm bloom must receive pollen that has been harvested earlier from male trees. Application of processed dry powdered pollen is done in two ways, only one of which requires Palmeros to go up into the treetops and walk on the branches. The other method is by "air-blowing" pollen.

Palmero ties fruit arm strands together

When pollinating, timing is everything

The time for pollinating is usually in February and March as the new fruit arm "pods" split and the resulting fruit blossoms reveal themselves. After a fruit blossom is open about five days or so and hasn't received at least a small amount of pollen, it's ability to accept pollen diminishes rapidly. For that reason, to ensure the best possible production, it is absolutely critical that pollen be applied at the appropriate time and in the most positive manner.

First the fruit strands are tied as bunches

In groves that use pollinating techniques developed by the early Medjool growing pioneers, the work is done manually. The first step in this method of pollinating is to locate the female flowers that are opening and remove the pod "sheath" (now split) from around the fruit "arms" by cutting it free. Next, the Palmero ties a string around the fruit "arm" collection of baby date-holding strands to hold them together for pollen reception. Then, using a hand-held shaker, the worker dusts the complete

A light "dusting" does the job

fruit arm with a light sprinkling of pollen. After dusting the fruit flower, the worker inserts a pollen flower "twig" into its center. The twigs are saved from the male pollen flowers after their pollen has been shaken loose at processing time. They have a small amount of remaining pollen on them so they are still useful.

A pollen flower twig is inserted

Insertion of the "twigs" provides a degree of additional insurance to further guarantee adequate pollination. During the pollination period it is sometimes necessary to return to the female trees more than once as additional fruit pods continue to open.

"Air-blowing" pollen is proving to work effectively

The other pollen application method is by air blower. Grower experimentation has indicated that this method works well and can be very efficient. Growers have devised their own hand-held compressed-air blowers that point and shoot a light blast of pollen into the fruit arms from a platform on a lifting vehicle. This makes it possible to move through the grove fast enough to complete the job before the fruit arms mature to the point where they will not adequately receive pollen. There are years when the weather causes the female fruit arms to bloom so fast that this method is the only way to get the job done before it is too late. It is much faster than the manual method and requires less labor…but more equipment.

Air-blowing gear

The pollen air-blowing equipment is carried on the work platform and consists of a portable air compressor producing up to 120 p.s.i. (pounds per square inch) of pressure...that forces air through a hand-held lightweight pipe. A "canister" of pollen is attached to the pipe in such a way that, as the air is released through the pipe, the pollen is sucked from the canister and forced under pressure to shoot out in a reasonably-controlled directional fashion.

Special attention for late blooming flowers

There are situations when the very last fruit arms to open are as late as early April. Then workers will make one last fast pass through the grove (usually using the air pollinator) to pollinate all the "late bloomers." Air-blowing works better when operating as a "clean-up" measure to pollinate late-blooming fruit flowers because it can be done much faster than the traditional hand method.

Growers must fabricate air-blowing equipment

Because of the very limited number of date growing operations in the U.S., this equipment is not available "off-the-shelf" and growers make their own versions for their own use. With some setups, the blowers will shoot too much or too little pollen with different "shots" so growers continue to work to improve their blowers to waste as little valuable pollen as possible. Some have experimented with constructing blower devices that accurately meter the amount of pollen that goes into each shot, a concept that can help to conserve a very limited supply.

Air-blower on lift pollinates late-blooming flowers

Bard Date Company, Yuma, Arizona

Among the production activities demanding the most care and precision is work involving the various management and manipulation of the fruit "arms." This worker is cutting a fruit "arm" center at its base to facilitate its removal for the purpose of extensive "thinning."

✒ruit arm de-centering, strand reduction & tying down

Training fruit "arms" and removing "centers"
Mid-March through late April

De-centering and thinning the number of cluster strings then training them to hang downward

This is the first step in thinning the dates. The procedure of this operation is to remove the center of the fruit arm bunch and leave only the outermost fruit arm strands. This work also makes possible other later necessary processes including individual strand date-thinning, strand-spreading with wire rings and "bagging." The large fruit arm strand centers contain from approximately forty to eighty-plus fruit strands. The arm centers are completely removed by cutting them out with a typical pruning shear. Next, remaining individual strands are removed from around the perimeter of the remaining arm in a balanced manner to arrive at the final number desired. Then the fruit arms are pulled down gently and tied with twine to lower branches in a position that begins their training in a downward drooping fashion. The tying downward is done to facilitate the fruit growing in a uniform controlled manner and in a location that will facilitate easier harvesting. Also, at this time, the remaining elements of the fruit flower arm pod sheathing are cut away and discarded. Pod sheath remnants are removed by cutting them free at their bases with a small hook-bladed knife.

Tying "de-centered" arms downward

Arms are meticulously "managed"

The arms must be carefully manipulated throughout the growing period so the mature fruit will have more room to grow freely and let more sunlight in to warm each individual date. Arm training also makes the date clusters more accessible for harvest. Typically, one man can usually complete the work of decentering and tying down one tree in about fifteen minutes.

Fruit clusters and stems are referred to as "arms"

The Medjool palm tree member that holds the fruit begins as a brownish "pod" about three inches plus in width by a couple of inches in thickness and grows from two to four feet plus in height. The pods are a very strong fiber with walls ranging to about three-sixteenths of an inch in thickness and encase the fruit member that has a stem about an inch thick by 2" wide. They grow vertically in the lower third of the palm treetop and a typical mature tree will produce from seventeen to twenty-five fruit pods. The actual fruit-bearing unit springs from the pod when the pod splits in two naturally to reveal it and is generally referred to by Palmeros as the fruit "arm."

"Clusters" hold thousands of baby dates

The fruit cluster is an assemblage of as many as 150 fruit "strands" (generally averaging about 100 to 130 on mature trees) of about an eighth of an inch or so in diameter. They vary from about 14 to 24 inches long and grow from the main fruit stem. The words "string" and "strand" are used interchangeably within the industry and mean the same thing. Each "strand" holds from approximately twenty to sixty (usually closer to 60) tiny fruit about 1/8 of an inch in diameter at the time of opening. When the pods are ready to flower and split in two (usually about early to middle March) to reveal the fruit "arms," the pod and arm/ strand assemblages

Removed fruit arm center
(about 50%+ of fruit arm total mass)

are in an approximate vertical position. When the female fruit arms open and bloom, their flowers of "strands" usually contain a total of from 6000 to 9000 baby fruit each. To achieve the goal of larger and more luscious fruit, radical thinning is done. That result is reached by greatly reducing the numbers of strands remaining on the arms, thereby allowing more nutrients to reach fewer units. The optimum number of ultimate fruit strands on each fruit flower arm allowed to grow to maturity is from 25 to 40 with each holding approximately 13 to 22 baby fruit…for an approximate total of 400 to 800 fruit per fruit arm.

Removing a fruit arm center

"Late-bloomers" are now pollinated

During de-centering and string-thinning, some very small fruit arm pods that weren't open earlier at pollination time may now have begun to open. They will be pollinated at this time in addition to de-centering and tying down if they are large enough and long enough. Also at this time, any additional smaller thorns that now present themselves will be removed. These are thorns that hadn't grown out by the time of the earlier "de-thorning" process. Typically, one man can usually complete the work of de-centering/ string-thinning of one tree in about thirty to forty minutes.

Pollinating a late bloom
with a simple hand-shaker

Removing old fronds

Fruit clusters receive additional "training" at this time

Fruit clusters grow heavier

These activities take place from late April to early June. Old lower fronds are now removed. At the same time, the fruit clusters are becoming quite heavy and need to be tied up (if necessary) to support their weight as well as to be repositioned for easier harvest. The fruit clusters will be tied securely to one and, sometimes, two nearby fronds as required.

Leaf (frond) removal

On mature date palms, approximately 100 leaves (fronds) are required to support and produce a normal date crop. The palm tree is a "monocotyledon," having a single stem (trunk) with a single terminal growth bud (shoot apex) at the top of the stem. Each year from ten to twenty-six leaves are produced depending upon the tree's age and health. They develop from the shoot apex and grow slowly for about four years before they emerge above the fiber. At this time, the "midrib" (rachis) elongates rapidly and the leaflets (pinnae) unfold. Attached to the sides near the base of each leaf is a fiber sheath which encircles the palm and protects the tender succulent tissue at the growing point (heart).

Fronds live from 3 to 7 years

Leaves (fronds) live from 3 to 7 years. Trunk growth takes place at the top of the palm and ranges from 8 to 30 inches per year. Because it does not have a "cambium" layer (as do other trees), the lower trunk does not enlarge as the palm grows.

Palmero removes old fronds

Fronds (branches) must be removed

Lower leaves must be removed each year as new growth springs from the top of the tree and the lower growth dies off. Since about fifteen or more new leaves are produced each year on a mature tree, no more than that same amount should be removed at any one time. If there are fewer than 100 total leaves on a tree, only the dead lower leaves should be removed. The number of leaves on a tree can be estimated by counting the number of tiers of leaves and multiplying by thirteen.

Frond removal processes are changing

In the past, most fronds were removed with a super sharp hand-axe. Two men could do a complete tree frond removal in about ten minutes. Times are changing rapidly. As costs of operation constantly go up, new methods are introduced to reduce man-hours expended. Many growers are switching to worker-operated hydraulically powered "shears" that can snip a frond from the tree very quickly reducing considerably the time required for the task. Although the equipment is expensive, it is proving to be "cost-effective" in most situations. Hydraulic cutting is a two-man operation. One man cuts while the other assists by pulling the fronds out of the clipper's way and directing them to open areas on the ground.

Fronds become mulch

Some growers grind the old fronds up with a mechanical device that works like a rototiller. It has very sharp rotating knives that reduce branches and also spent fruit arms to smaller elements that will then decompose into mulch. Other growers feed their old fronds through a motorized chipper and then spread them out in the grove to become mulch.

The task of "thinning"

Mid April through Early June

Very large dates are the ultimate goal

The diameter of any ordinary tree's root system is mirrored by a similar (approximate) diameter of its above ground foliage. This formula applies to the date palm as well. Its ability to grow dates is a reflection of the size of its root and palm leaf spread. In the past, Medjool palms were usually planted thirty feet apart in rows thirty feet apart. Some plantings have become more dense in recent years. The spacing must allow the Medjool palm to grow to its maximum producing size and provide ample room for the roots to acquire adequate nutrients as well as enough open space for each tree to have full access to growth-enhancing sunlight. Over time, experimentation has shown this approximate tree spacing to be the optimum for achieving the highest date production.

Radical "thinning" is the key to very large fruit

If the date palm fruit clusters are left to grow in a completely natural state, they tend to greatly over-produce which results in much smaller individual dates and fewer date arm "flowers" in following years. It has been determined by study that the number of fruit bunches on a mature tree is more or less directly proportional to the total number of leaves it has. The average mature Medjool palm with 100 branches will produce enough nutrition to support up to 25 fruit-producing pods. The pods produce stems with fruit flowers on their ends that become the hanging date clusters. These production numbers assume that enough tree spacing has been provided to supply the maximum nutrients possible.

Dates must grow larger before thinning

Thinning is the most exacting task

To achieve the ultimate date size, the individual fruit "strands" on the fruit "arms" are manually "thinned." Thinning of the infant fruit allows greatly increased food value to flow to the remaining fruit. As a result, the remaining fruit grow to as much as three times the size of ordinary dates which is what makes the massive juicy and soft Medjool dates so highly prized. Growers must be very careful about exactly how they thin their dates. If not enough are removed in the thinning process, the quantity produced will increase dramatically but the quality and size will diminish. Because the Medjool date can be forced by thinning to grow to surprisingly large dimensions and very high quality when left to grow open and uninhibited, that is the usual method of operation employed by most Medjool growers. The time for fruit string thinning is usually May and June. Thinning is the most exacting, important, time-consuming and tedious job of all the many tasks in a date grove.

Initially, "arms" hold thousands of baby dates

When it blossoms, the Medjool date fruit arm with its many fruit strings contains from 6,000 to as high as 9,000 tiny baby dates. Growers have learned the optimum number of baby dates left to grow to maturity is from 400 to 800 per fruit arm. This allows the remaining dates to grow to an exceptionally large size and that number of dates on a cluster works out to about 8 to 12 pounds. The only way to accomplish this goal is to reduce the number of fruit strands

Unthinned fruit strands illustrate the high density of initial infant date growth

first (which was done during the earlier fruit arm "de-centering" process) and then remove individual baby dates from each arm _manually_, an extremely labor-intensive task. Approximately eight or nine of every ten baby dates that first appear are removed by hand in the thinning process. Many other varieties of dates grow to maturity without the necessity of this demanding effort. The resultant benefit for them is not worth the immensity of the task. Even with dramatic thinning, the remaining Medjool dates are so voluminous that they can be very crowded within their clusters.

Thinning from a work platform

Tall mature trees require use of a lift vehicle

A lift vehicle raises workers in the lift platform to the bottom of the palm spread and then slowly eases it up a bit more until it touches (and raises) the branches slightly. This allows the work platform to be as far up as it can possibly go without damaging the tree and gives the workers a solid platform from which to accomplish their task. The work platforms have metal safety railings on which walking planks (Palmeros call them "crossboards") can be placed at different levels so workers can easily reach the date arm clusters that hang down.

Heavy with dates - thousands to be thinned out by hand

Baby dates are removed one at a time

The removal of the majority of the growing baby date buds "one-at-a-time" from the fruit strands demands tremendous skill, dexterity, patience and experience on the part of workers. Most growers will leave a space of approximately one inch to one and one-quarter inches between buds leaving a final number of 13 to 22 buds per strand (of which there are from 25 to 40 strands per fruit arm). That means removing well over three fourths of the baby fruit! Each fruit arm of the usual approximate number of two-dozen total on mature trees must be painstakingly thinned individually. The Palmero carefully unties the fruit arm (it was tied during "de-centering" to help "train" it to stay together and bend downward) and holds the individual fruit arm "strings" apart...then with his fingers removes the great majority of the baby dates from each string. The goal is to finish the process with each fruit "arm" holding approximately four to eight hundred dates. When the middle of the thinning process has been reached, the size of the growing dates ranges from a quarter-inch to a half-inch in diameter.

Thinning is a painstaking process

Performing this meticulous process at top speed, a Palmero can thin only about six to eight trees per day under best conditions. The fastest workers can thin a tree an hour. A more likely average is an hour and a half per tree varying with the height of the tree, number of fruit clusters on a tree and how thick the

Young date cluster thinned

fruit is within the clusters. If the trees are mature and the pollination operation was highly effective, the fruit strings will be extremely congested and this condition understandably increases the thinning time. When finished thinning, approximately 90% of the fruit buds will have been removed.

Fruit arms now receive further "training"

After thinning, the fruit arm strings are left to hang loose to be open and ready for the next major task which will be inserting "spreader rings" and "bagging." At this stage, the fruit arm stem now might be moved a bit and tied to lower branches to continue training downward. This will hold it in a location that will allow it room to grow free and offer easier access for harvesting. If a fruit arm is growing in a way that might cause physical interference with other branches or other fruit arms, it will be moved to one side (to be in the clear) and tied in place. If fruit arms are allowed to rub against other fruit arms or branches, the result will be bruised and damaged or lost fruit so it is very important to separate them from each other as well as from branches. The goal of the training process is to have fruit arms growing in as free a space as possible and hanging straight downward to provide the optimum position for harvest.

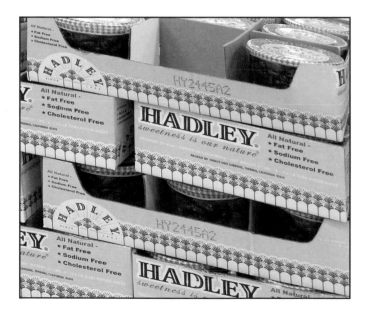

Hadley Orchards, Cabizon and Thermal, California

Ringing and bagging begins

Bundle of 8" diameter "spreader" rings
(used to separate individual date strands)

"Ringing" and "Bagging"

Ringing and bagging to separate and protect

About the end of June and running into July, growers begin the simutaneous processes of "ringing" the interior of each date cluster and "bagging" (placing a cover completely over each cluster).

Carefully, "bags" are placed around each date arm cluster

"Rings" are now inserted into the cluster centers

A wire "spreader" ring is inserted inside each date strand cluster. The "ring" is a circle of galvanized wire about 8 inches in diameter with kinks formed into it. It is inserted inside the center of the date cluster "strands" to separate them and hold them apart so the fruit won't touch. By this time, dates will have grown large enough (about a quarter to a third of their ultimate size) to hold the rings in place . Ringing reduces fruit loss by dates becoming scarred with abrasion from neighboring dates. The separation of strands also allows dates to receive more ventilation and better atmospheric conditions to grow as large as possible.

Rings are custom made

The rings are custom made by machines specifically designed by the date growers for this purpose. Infrequently needed, very few growers or members of date-growing cooperatives may have such a ring-making machine but those who do often generously loan it to other growers when needs arise.

Bags are now afixed

Cloth "bags" are put on at this time. Bags act first to protect the ripening fruit from inclement weather. Later, when the dates are ripe, the bags serve as a harvesting assist tool. The bags are a fabricated cloth cylinder about two feet wide if laid flat and approximately four feet long. They are a "mesh" that resembles cheese cloth and are moderately resistant to water intrusion. With rings inserted into their centers, the date clusters are now expanded enough that it takes a fair amount of effort to get the bags over the entire clusters. To hold them firmly in place, the bags are wrapped around the date bunch and tied together in a "gather" at the fruit arm top. Some growers will leave a few date clusters un-bagged on perimeter trees or on trees that have some inferior fruit to act as "decoys" to lure birds away from the best fruit. Growers occasionally find bats in the bags but they are not a major problem. Initially the "bags" are left open at the bottom as the fruit ripens to allow as much aeration for as long as possible.

Thousands of clusters must be bagged

Initially, all bags are left open at bottom

A grove after bagging with bottoms now closed

As fruit ripens, the bag bottoms are closed

About the last week of August or the first week of September, depending on weather variables, perhaps 10 days or so prior to harvest when brown tips begin to show on the bottoms of the dates, concern grows that ripe dates will separate and begin to fall so the bags are closed at their bottoms to prevent now-ripe fruit from dropping to the ground. This is a separate operation. Bag-closing time can vary as much as a month or so. Closing bags also provides a modest amount of protection from birds and insects. Usually, birds don't attempt to fly up into bags from below. Still, birds cause some damage by attempting to pick through the bags from the outside to get at the fruit. There are some situations where birds can cause as much as five percent crop damage. Some insects attempt to get to the date clusters by working their way through the bag top and then down into the cluster along the fruit arm from above. That is the reason the bags are tied very tightly at the top. When fruit ripens to the point that some dates begin breaking loose and falling and enough dates can be seen accumulating in the bottom of the bags, the Palmeros will initiate the first "picking" and the harvest begins.

New bags are made, old bags are repaired

Paper bags are used on some date species

The reason paper bags can be used on other date species is because their main purpose is to protect those dates from inclement weather. In most Medjool groves, the cloth bags protect the clusters from insects and weather, maintain warmth and, in addition, the cloth bags play a major part in the harvesting process. Bags are a significant item when one realizes the quantity necessary. The average mature Medjool palm will produce from 17 to 25 fruit arms that must be covered. A grove of 1000 trees possibly could require as many as 25,000 bags, a typical number being from 15,000 to 20,000.

Bags are a costly expense!

Because the bags accumulate dust and date sugar residue and the cloth pores become plugged, ventilation potential reduces so growers wash the bags yearly or bi-yearly. If the bags do not allow adequate ventilation, when moisture occurs mold can grow and fruit can be ruined. Thus, it is imperative to have clean bags. An additional benefit of the bag washing is the elimination of potential for fruit contamination by foreign material in the following season. At season's end, the bags are washed as necessary, tied in bundles, usually of 25, then carefully stacked and stored for next season's use. The custom-made bags are a substantial expense so handling them with care and saving them for further use is a necessity. Each year approximately ten to fifteen percent of the bags must be replaced due to wear and damage.

Martha's Gardens, Yuma, Arizona

Years of experimentation with bags

In times past, as growers were experimenting with finding just the right type of bag, they tried using bags with a very tight weave to guard against rainfall and insects. They also tested waxing the top half of the bags to inhibit moisture penetration. After long years of experimentation with tight weaves and waxed tops, growers concluded the dates were not getting enough air circulation for adequate growth and those bags also led to damage by mold due to trapped moisture. They finally settled on fine-open-mesh cloth because of the dates' important need for adequate aeration. Although it is not the perfect solution and some crop loss results, the soft mesh bags have proven to be the best protection.

Bagging is a substantial "work-effort"

Under the best conditions, a two-man team can ring and bag from 800 to 900 fruit clusters per day. Such production requires that strings, rings and bags all be in place on the lifting equipment and ready to go in early morning. In a grove of 1000 trees, a total work effort of approximately 25 to 30 or so "man-days" of work are needed to complete the grove task of "ringing and bagging."

Harvesters begin first "picking"

Harvesting Medjools

Many date variety clusters are removed as a unit

Throughout the date producing world, a large quantity of dates are harvested simply by waiting until the greatest number of the dates are ripe, then cutting off the entire date arm cluster, lowering it to the ground and removing the dates later at a processing facility...*efficient*...and *quick*!

No such easy harvest with MEDJOOLS...
There is no removing clusters *whole* with Medjool dates!

Medjool dates are far too valuable a commodity to be treated in such a casual manner. There are important considerations in the Medjool date growth pattern that make them a much more labor-intensive date to harvest. Medjools ripen at different times over a four to six-week-long period. This precludes simply cutting the clusters off whole. As a result, Medjools receive a much higher degree of care in every step of the growing process which ultimately assures that the absolute finest condition and quality of date reaches the consumer.

Each date cluster holds hundreds of dates

Medjool date palms begin to bear fruit in 3 to 5 years and reach full bearing age between 10 and 15 years. Their dates begin to ripen about the last of

August or early September. The harvesting process usually runs from about the 1st of September through the end of the month, though sometimes a week earlier or later. As the dates ripen, it is necessary to make sure the heavy bunches are adequately tied to branches above to reduce the possibility of the arms breaking off due to the great weight. Tying the fruit arms also makes possible positioning them all at approximately even levels which then makes harvest much easier. At their heaviest, the date arm bunches can weigh 10 to 20 pounds. Most groves plan on three separate pickings.

Bag held open to illustrate ripening progression

Yields as high as 200 pounds per tree

Averaging 17 to 22 fruit arms, a mature tree yield can top 200 pounds. Although younger trees produce less, they are much easier to harvest because of their lower height. Harvesting mature trees that reach forty feet and higher requires the use of high-lift equipment and more manpower plus moving the equipment from high tree to high tree is more time consuming.

Bags are shaken gently and dates let down softly into receiving basket

Medjools are allowed to ripen on the tree

Medjool dates ripen on the tree because that is the very best environment for curing dates due to natural forces and high ambient temperatures where they are grown. Although it is possible to effect a degree of curing after picking, it is not easy to do and does not always achieve a result worth the considerable

Full baskets are lowered to Palmeros below

effort involved. Therefore, it is best to let them ripen completely on the tree... just one of the many reasons Medjools are a complicated crop to grow.

Medjools are not actually "picked"

Medjool dates are not actually "picked" but that is the word typically used by Palmeros in the Medjool growing community. The correct term is "harvesting." Human hands actually touch very few Medjool dates during harvest keeping the fruit as thoroughly clean as absolutely possible.

Special harvesting equipment is required

A typical harvesting crew will consist of two platform lifting vehicles with drivers and two or three Palmeros who go up into the trees to remove ripe dates plus at least one "runner" on the ground to get the fruit from the trees to the transport vehicle (usually a tractor pulling a low flatbed trailer). To facilitate easier access for the harvesting process in the trees, "walking planks" can be placed across

Ground level workers receive dates

(and rest on) the upper part of the high-work platform railings. When the platform is raised up, it then can be pressed up against the bottom branches to a point where the harvesters can reach the date bunches while still standing on a relatively stable foundation.

Medjools are picked three or more times

Medjool trees are generally harvested three different times per season and, in unusual circumstances of weather or ripening situations, sometimes more…depending upon how those (or other elements) may affect the ripening process. The dates are picked as they ripen which yields the largest sizes possible. The time for the first picking is determined by viewing the bags from the outside. Upon seeing some discoloration on the bag sides and bottoms caused by ripe fruit touching them, the harvesting begins. Also, some early ripe fruit (and some jarred loose by wind) will fall from the arm cluster and accumulate in the bottom of the bag indicating it is the time to begin harvesting.

Bags are retied after first "picking"

Harvesting from ground level comprises only 10% of the activity...
most of this work is done high in the treetops

The first harvesting can be light or heavy

The first harvesting is usually light producing roughly five to fifteen percent of the total harvest. There will be times when the first picking is much
heavier due to unseasonal weather preceding (and hastening) the ripening
process. Because the fruit clusters are so congested within their bags, it is not
very easy to actually "pick" the dates individually. Although some growers
do attempt to pick somewhat selectively, most use a bag-shaking method of
harvest. The process involves hanging large round shallow canvas baskets be-
neath each fruit arm bag and fastening them above the bag with three cords
that have quick-disconnect hooks. The fruit arm bags are then grasped and
very gently shaken to break free all additional dates now ripe enough to fall
easily into the bottom of the tied bag, then into the baskets.

After last picking, bags are removed

Great care must be applied at every step

This must be done with exceptional care to reduce the possibility of freeing dates from the fruit arm which are not ripe enough for the market. If detached, these unripened dates would become a complete loss. It is not possible to avoid shaking some unripe fruit into the bags so some waste simply is tolerated although Palmeros apply a very light touch to the shaking process. After the bag has been shaken adequately, the bag bottom is untied and the loose dates allowed to drop a very short distance into the basket. The basket is then moved to the next fruit arm cluster in line and the task repeated. When the basket is loaded (usually with from ten to twenty or so pounds of fruit), it is tied to a sturdy rope and sent down to a "runner" on the ground. A tree is usually "picked" in ten to fifteen minutes. Moving from tree to tree consumes a few more minutes per tree.

The harvesting ("pickings") continue

After the first date removal, the date arm enclosing bag is retied at the bottom to await the second picking. The harvesters re-tie the bottoms of the bags after each of the first three pickings. The second picking will usually be heavier and the third picking generally quite a bit lighter. At the final picking, all remaining ripe fruit is generally removed and the bags removed as well. There will be some unripe dates remaining but not enough to warrant another full-blown round of picking. Even though these dates may ripen later, most will be left behind to go along with the fruit arms when they are removed and turned into mulch. Some growers remove the spent fruit arms during the last "picking." Others remove them later in a separate operation.

Dates are handled with "kid gloves"

In some harvesting situations, the "runner" on the ground handles the fruit and also operates the fruit-hauling vehicle. If there is too much fruit for one person to handle efficiently, there may be an additional worker to operate the hauling equipment. The ground worker empties the baskets into thin stacking trays on the trailer and returns the basket by rope to the harvester. Because Medjools are extremely delicate and soft, they crush easily so they are spread out evenly just one date deep in the tray to eliminate fruit damage. The 2'x2' trays are thin (about an inch and a half deep) and constructed with an arrangement on the corners that allows separation with an airspace between trays when stacked.

Trays are transitioning to other materials

The thin trays offer the possibility of later individual date exposure for further air ripening and sun drying if required, as well as reducing the possibility of date damage due to crushing by too much weight. Originally, only wood was used for the trays but, as times change, growers are on the lookout for other possible materials to do the job such as recycled materials or other environmental-friendly methods.

On to the packing plant

Finally, the dates are delivered to the processing facility where they will be sorted, graded and packaged into what many consider to be the finest "natural confection" of all.

Trays ready for the packing plant

Dates dry for a short period to "firm" and ripen

𝒯he 𝒫acking 𝒽ouse

Grading and packaging

Occasional Giants!

Medjools are graded carefully for quality

After harvest, usually from late August through October, the fruit is taken to a processing and packaging facility where it is carefully graded and packaged. The grading operations are just one more demanding labor element of Medjool production that differs from how most ordinary dates are processed. Because the Medjool date is the softest variety, it is essential that it be handled delicately to avoid skin damage and bruising and to maintain the utmost high quality. Of course, bruised skin is only a cosmetic blemish and certainly doesn't affect the taste but all Medjool growers work diligently to prevent anything that might adversely affect the high quality and best possible presentation of their product. The dates are sent over a moving belt conveyor past grading stations where skilled packing house graders sort and separate

Quality processing is a very busy time

them according to condition, size and quality. Because there are so few Medjool growers in the U.S., most have been forced to establish their own grading, packing and storage facilities. Even some of the smaller growers have designed and built their own packing houses and storage operations to guarantee maximum quality control.

There are a number of date grades

Various growers grade their dates somewhat differently from each other according to their individual plant operations and marketing situations. Some growers will grade as many as seven or eight categories including Jumbo, Select, Large, Fancy, Extra Fancy and Choice. Some may grade as few as two or three classes of fruit. As weather and growing conditions vary from year to year, one grade may be produced in much larger quantities than it was in a previous year.

Growers give Mother Nature an assist

Any dates that may have come to the packing house in a not-quite-fully-ripe condition will be separated and placed on trays that will be

Dates are sorted by size

exposed to warmth in an effort to complete the ripening process. As a rule, this process is only marginally successful.

Some years the best quality grades are produced in relatively large numbers while other years generate a preponderance of lower-quality dates. The growers chuckle a bit when they say, "If we knew exactly why such natural phenomena occur, we would produce only the very best every time...but, unfortunately, that is not the case." Even the most knowledgable and learned horticulturists can be at a loss to explain how some years will differ significantly in date production and quality. The mysteries of Mother Nature remain an enigma that date growers continually try to decipher.

Medjools packaged for distribution in five-kilo retail display cartons

Medjools are distributed worldwide

The majority of the harvest is sent far and wide to retail grocery stores. Many Medjools are packaged in five-kilo cartons that double as display units in the stores. As more groves are planted and more fruit is becoming available, the network of stores selling the Medjool date is expanding rapidly.

Medjool dates make an unforgettable gift

A significant portion of the harvest goes into "Gift" packaging. After grading, "Gift" dates are moved immediately into cold storage. Later, they will be repackaged into assorted appropriately-sized and gift-styled containers. A surprising quantity of the Medjool harvest is purchased by knowledgable and discriminating buyers and sent to friends, relatives and business contacts. Medjools make an outstanding gift because of their great taste, texture, uniqueness and relative rarity along with their inherent ability to retain a high degree of freshness after long periods either refrigerated or frozen.

✒ertilizer

Natural fertilizers are applied

Palm roots go deep and wide

Date palms develop very deep and spreading root systems capable of foraging widely for water and nutrients. Maximum growth and fruiting will continue for many years in a moderately fertile soil with a minimum of additional fertilization. However, like most agricultural plants, extra fertilizer judiciously applied can return substantial benefits.

Medjool growers are going "organic"

In earlier times, urea fertilizers high in nitrogen were applied two or three times a year to many date groves. Public demand has changed that growing model significantly in recent years. Most growers now are growing organically eliminating the use of most, if not all, chemical fertilizers. To qualify as "organic", they cannot apply any "herbicides" or "pesticides." When growers go "organic", after three years of growing in such a fashion they can be government-certified as "organic."

"Organic" is not a new concept

People have been growing organically for thousands of years. Modern farmers are learning all they can as rapidly as they can to apply these age-old principles to produce a better product. Medjool date growers are in the forefront of this movement.

The government has set organic standards

The federal government adopted standards in 2002 for organic certification. The rules prohibit the application of synthetic pesticides and fertilizers, genetic engineering, irradiation and the addition of sewage sludge in growing operations. As of the 21st century, not all date growers are yet so certified but many are working in that direction. Some have sections that are organic and other sections that are not. There are growers who say that when comparing two groves in close proximity, one organic and the other not, it can be next to impossible to tell any difference in the trees or the fruit.

Ꮚedjool Ɗates ᐱre ᏔighIу Ꮤutritious

Medjools contain many vitamins and minerals

Medjool dates are rich in flavor and sweet to the level of many candies. They also possess tremendous nutritional value for their size, add much to an everyday diet and make an outstanding alternative food for those who are health-conscious. Dates are full of energy-producing carbohydrates, protein, magnesium, niacin, fiber, vitamin B complex and other minerals while they have no sodium, no fat and no cholesterol. Dates contain more potassium than bananas, are high in sugar content and rich in iron. Each fresh Medjool date is only about 40 calories and studies suggest that regular inclusion of dates in a diet may lead to reduced rates of some illnesses.

Medjools' amazing nutritional statistics

Actual figures for vitamins and food value indicate a serving of three to four large dates (United States Recommended Daily Allowance) has approximately 120 calories, 240 milligrams of potassium (7% USRDA), 31 grams of carbohydrate (10% USRDA) including 3 grams of dietary fiber (14% USRDA) and 29 grams of natural sugar plus 1 gram of protein. Such a serving of dates also has 2% of the USRDA for calcium, thiamin, niacin, iron, riboflavin, Vitamin B6, phosphorus and 4% of the USRDA for magnesium.

Medjools store well in refrigerator or freezer

Medjool dates can be stored at room temperature up to a month and, in optimum atmospheric conditions, somewhat longer. Much longer life is possible if kept in the refrigerator...as long as six months. As with any food held in the refrigerator, dates will become drier with time. However, they can be quickly re-humidified and softened by simply putting them in a covered pan with a small amount of water in the bottom then placing them in an oven at low temperature for twenty minutes or so. If prepared by "vacuum" packing in their natural state, Medjools can be frozen and retain their quality for a year and even longer...depending on the percentage of air removed...and if held below 10 degrees fahrenheit. Their high sugar content keeps them from freezing hard. They truly are a wonderful-tasting and highly-convenient food.

Section III

This assemblage of recipes utilizes the legendary fruit of the desert long revered as the "staff of life" in areas of the world where civilization has its roots.

Because of its outstanding natural sweetness and flavor, the date has continued to be a favored feature in the formulas of countless culinary concoctions throughout world history.

Many of the recipes in this grouping were conceived in primitive smoky sculleries of the distant past and evolved over centuries of imaginative innovation.

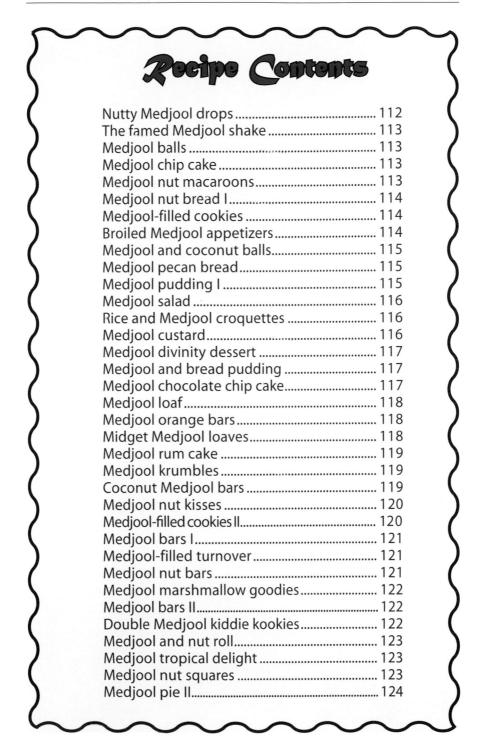

Recipe Contents

Recipe Contents

Recipe Contents

An All-Time Classic...
Nutty Medjool Drops

3/4 cup Medjools	1 egg
2 cups flour	1/2 cup pecans
1 cup sour cream	1/2 teaspoon salt
1 cup brown sugar	1/2 tsp. cream of tartar

1/2 teaspoon baking soda

Cream brown sugar and sour cream. Add egg. Mix soda, cream of tartar, salt and flour. Add to creamed mixture. Stir in Medjools and pecans. Drop by table-spoonfuls onto greased baking sheet. Bake at 350° for 10 to 12 minutes or until slightly browned.

The Famed Medjool

"Date Shake"

2 LARGE scoops hard vanilla ice cream
1/3 cup pitted and finely chopped Medjools
1/3 cup milk
Blend in ordinary blender or stir in milkshake machine
until well blended, but...*very thick.*

Medjool balls

1 pound Medjools, pitted
Chop them in a processor with 1 cup pecan meats
Add: 1/4 teaspoon salt
Shape them as candy into tiny balls
Roll them in powdered sugar

Medjool chip cake

1 cup boiling water   1 cup sugar   1 cup chopped Medjools
1 teaspoon vanilla   2 eggs   1/2 cup margarine
1 3/4 cups flour   1/2 tsp salt   1 teaspoon soda
    1 tsp baking powder   1 tsp unsweetened cocoa

Topping:

3/4 cup chopped nuts   1 cup mini semi-sweet chocolate chips

Pour water over Medjools; cool to lukewarm. Preheat oven to 350 degrees. Grease and flour 13x9 inch pan. In large bowl, cream sugar, margarine, vanilla and eggs. Add flour, soda, baking powder, cocoa, salt and date mixture. Beat 2 minutes at medium speed. Pour into prepared pan. Sprinkle nuts and chocolate chips over batter in pan. Bake 35 to 40 minutes until toothpick inserted in center comes out clean.

Compliments of Vicky Daniels

Medjool nut macaroons

1 cup flaked or shredded coconut   1 cup chopped nuts
1 cup pitted, chopped Medjools   1 teaspoon vanilla
2/3 cup sweetened condensed milk (not evaporated)

Mix together all ingredients. Shape into balls and place about 1" apart on greased baking sheet. Bake in moderate oven at 350° 10 to 12 minutes until golden brown. Remove cookies and cool on racks.

Medjool nut bread I

1/3 c up butter or margarine	¾ cup sugar
¾ cup buttermilk or sour milk	2 eggs
2 cups unsifted flour	½ teaspoon soda
½ teaspoon baking powder	½ teaspoon salt
½ cup chopped Medjools	½ cup chopped nuts

Preheat oven to 350°. In mixing bowl, cream butter and sugar. Beat in eggs; blend in buttermilk. Stir in remaining ingredients just until blended. Pour into 5x9-inch loaf pan, greased on bottom only. Bake 50 to 55 minutes or until toothpick inserted in center comes out clean.

Medjool-filled cookies

1 cup shortening	2 eggs
2 cups brown sugar	½ cup warm water
1½ teaspoons soda	1 teaspoon vanilla
½ teaspoon salt	4 cups sifted flour

Cream shortening and sugar. Add eggs one at a time. Mix well. Dissolve soda in warm water and add to mixture. Add vanilla. Add flour, etc. Bake at 375° 10 minutes. Drop by spoonfuls, make dent for filling and cover with cookie dough.

Filling:

1 cup chopped Medjool dates	½ cup sugar
1 cup water	2 tablespoons flour
1 tablespoon lemon juice	½ cup chopped nut meats

Cook filling until thick.

Broiled Medjool appetizers

¾ cup brown sugar	½ cup water
1/3 cup lemon juice	¼ cup cider vinegar
1 teaspoon grated orange rind	½ teaspoon cinnamon
¼ teaspoon nutmeg	1/8 teaspoon salt
1 8-ounce package Medjool dates	¾ pound sliced bacon

Combine sugar, water, lemon juice, vinegar, orange rind, spices and salt in a saucepan. Place over moderate heat, stirring until sugar dissolves. Bring mixture to a boil: reduce heat and simmer 5 minutes. Pit and place Medjools in small glass bowl. Pour spice mixture over dates; cover and place on rack to cool. Refrigerate at least 24 hours to allow flavors to blend. Drain dates; wrap each date in ½ slice bacon and skewer with toothpick. Broil until bacon is crisp. Serves 6.

Medjool and coconut balls

2 tablespoons butter	1 cup sugar
2 eggs, beaten	1 cup chopped Medjool dates
½ cup chopped nuts	2 cups crisp rice cereal
1 cup flaked coconut	

Put butter in heavy skillet with sugar, eggs and Medjools. Cook over medium-low heat, stirring constantly until mixture leaves sides of skillet (mixture burns easily). Remove from heat; add nuts and cereal. Shape in 1" balls with hands; roll in coconut. Store in tightly-covered container (cookies are good keepers). Yield: 38

Medjool pecan bread

2 cups boiling water	3 teaspoons baking soda
1 ¼ cups dates, in quarters	3/4 cup brown sugar
2 teaspoons salt	4 tablespoons butter
4 eggs	4 cups sifted flour
1 teaspoon baking powder	1 cp finely chopped pecans

In a bowl combine quartered dates, soda and water. Let stand 15 minutes. In another bowl combine butter, salt and sugar. Beat in eggs vigorously until a smooth batter is formed. Fold in date mixture. In a separate bowl combine flour and baking powder. Add to Medjool-flour mixture blending thoroughly. Blend in nuts...1/4 cup at a time. Pour batter into a greased bread pan. Let stand 15 minutes. Bake at 325° for 1½ hours until brown. Use a toothpick to check for doneness. Remove pan to a rack. Let stand 5 minutes. Invert pan onto a breadboard or platter. Cut into 1-inch thick slices. Serve plain or top with peach ice cream. Yield: 1 loaf.

Medjool pudding I

4 egg whites	¼ cup (heaping) sugar
3 tablespoons flour	Pinch of salt
1 teaspoon baking powder	2 cups chopped Medjool dates
1 cup chopped nuts	1 teaspoon vanilla extract

Beat egg whites in bowl until stiff peaks form. Add sugar gradually, beating until very stiff peaks form. Sift flour, salt and baking powder together into bowl. Fold gently into egg whites. Fold in remaining ingredients. Pour into greased 9x11-inch baking pan. Bake at 325° for 1 hour or until set. Serve warm or cold. Yield: 6 to 8 servings.

Medjool salad

3 (3 oz.) pkgs. cream cheese
1 tablespoon lemon juice
1 cup chopped Medjool dates
½ cup chopped nuts

¼ cup maple syrup
1 (8 oz.) can crushed
   pineapple (drained)
1 cup cool whip

Beat cream cheese until creamy. Add maple syrup, lemon juice, drained crushed pineapple, chopped Medjools and chopped nuts. Fold in cool whip. Keep refrigerated.

Rice and Medjool croquettes

3/4 cup Medjools, pitted and cut fine
1 cup rice
1/4 teaspoon salt
Grated rind of 1/2 lemon
1 egg

2 cups milk
1 egg, well beaten
2 tablespoons sugar
Bread crumbs

Cook rice, milk and salt in double boiler until the rice is tender and milk is absorbed. Remove from heat, add the beaten egg, lemon rind, sugar and dates and mix well. Spread mixture on a large platter to cool; then shape into croquettes. Dip in well-beaten egg and then in bread crumbs. Fry in deep hot oil until golden brown. Drain on unglazed paper and serve with currant jelly.

Medjool custard

1 cup Medjools, pitted and chopped
6 tablespoons flour
4 tablespoons butter

½ cup sugar
3 egg yolks
1 pint milk

Dash of vanilla

In the top of a double boiler, blend 3 tablespoons of butter and flour. Add milk. Heat, stirring constantly until thickened. Add sugar and egg yolks. Cook over hot water 10 minutes. Flavor with vanilla. As it is nearing completion, stir in 1 teaspoon of butter to make it smooth. Put the Medjools in a dish. Pour the custard over them.

The pudding may be served in sherbet glasses by placing a portion of Medjools in each and pouring the custard over. This should be served either hot or cold with cream.

Medjool divinity dessert

1 cup Medjool dates pitted and chopped (coarse)

6 macaroons or ½ cup coconut	1 package gelatin
¼ cup cold water	1 cup hot water
½ cup sugar	¼ teaspoon salt
½ cup orange juice	1 tablespoon lemon juice
½ cup cream (whipped)	Optional: Chopped nuts

Pour cold water in bowl and sprinkle gelatin on top of water. Add sugar, salt and hot water and stir until dissolved. Add fruit juice and cool. When it begins to thicken, add Medjools, crushed macaroons or coconut. Fold in whipped cream. Pile into glasses and chill. Sprinkle tops with nuts just before serving.

Medjool and bread pudding

½ cup chopped Medjool dates	½ cup chopped nuts
½ cup flour	¼ teaspoon salt
1 teaspoon baking powder	¾ cup honey
½ cup whl wheat bread crumbs	2 eggs, beaten

Dust Medjools and nuts with a small amount of flour. Sift remaining flour with salt and baking powder. Add honey to eggs in a fine stream in bowl, beating constantly. Add crumbs, dry ingredients, Medjools and nuts; mix well. Pour into greased 9-inch baking dish. Bake at 350° for 20 minutes. Serve warm. 6 servings.

Medjool chocolate chip cake

1 cup chopped Medjools	1 cup margarine, softened
1½ cup flour	3 tablespoons baking cocoa
1 teaspoon soda	½ teaspoon salt
2 eggs, slightly beaten	1 tsp vanilla extract
½ cup chocolate chips	1/4 cp finely chpd pecans
1 cup sugar	1 cup brown sugar

Put Medjools and 1 cup boiling water in bowl. Add flour, cocoa, margarine, soda and salt; mix well. Stir in sugar, eggs and vanilla. Pour into greased and floured 9x13-inch cake pan. Sprinkle with mixture of nuts, chocolate chips and brown sugar. Bake at 350° for 40 minutes or until cake tests done. Cool in pan. Makes 18

Medjool loaf

2 cups pitted chopped Medjool dates
1 cup chopped walnuts
1/3 pound marshmallows chopped
1 teaspoon vanilla

Crush 1/3 pound
graham crackers
1 cup heavy cream
whipped stiff

Combine ½ the cracker crumbs with the Medjools, marshmallows, nuts and whipped cream. Shape them into a roll. Roll it in the remaining cracker crumbs. Chill the roll for 12 hours. Serve it cut into slices with cream or whipped cream.

Medjool orange bars

½ cup chopped Medjool dates.
½ cup brown sugar
1 teaspoon grated orange peel
½ teaspoon baking powder
¼ cup milk

¼ cup butter
1 egg
1 cup flour
½ teaspoon soda
¼ cup orange juice

¼ cup chopped walnuts

Cream butter and brown sugar until fluffy. Add egg and grated orange peel; beat well. Sift together flour, baking powder and soda; add to creamed mixture. Stir in milk, orange juice, walnuts and Medjools. Spread in greased 11x7x1½ inch pan. Bake at 350° for 25 minutes. Cool; sprinkle with confectioner's sugar. Yield: 24

Midget Medjool loaves

8 ounces pitted, chopped Medjools (1 1/2 cups)
1 egg
1/2 cup orange juice
1 1/2 cup sugar
1 tspoon baking soda
2 tbsps shortening

1 tblspoon grated orange peel
2 cups flour
1 teaspoon baking powder
1/2 teaspoon salt
1/2 cup chopped walnuts

Pour ½ cup boiling water over Medjools and 2 tablespoons shortening; cool to room temperature. Add orange peel and orange juice. Stir in 1 beaten egg. Sift together flour, sugar, baking powder, baking soda and salt; add to mixture; stir until mixed. Stir in ½ cup chopped walnuts. Turn into 4 greased 4 ½ x 2 ¾ x 2" loaf pans. Bake at 325 degrees for 40 to 45 minutes. Remove from pans; cool. Wrap and store overnight. Optional; Bake in 9x5x3" loaf pan for 1 hour.

Medjool rum cake

1 lb. Medjools, chopped	1 1/2 cups brown sugar, packed
1 teaspoon baking soda	3/4 cup butter
1 lb. chopped walnuts	1 cup boiling water
3/4 teaspoon salt	2 tablespoons rum
3 eggs, beaten	2 1/4 cups flour

Cream butter and sugar; beat in chopped Medjools and nuts. Mix baking soda with water and pour into mixture. Add beaten eggs. Add flour and salt, beating until smooth, then stir in rum. Pour into a greased and floured 13 x 9 x 2-inch pan and bake at 350° 25 minutes or until toothpick inserted in center comes out clean. Remove from pan and apply Rum Glaze.

Rum Glaze:
Mix together 1 cup powdered sugar with 1 tablespoon rum; add water to desired consistency. Spread over warm cake.

Medjool krumbles

1½ cups flour	1¾ cups rolled oats
1 teaspoon soda	¾ cup butter
1 cup brown sugar	1 teaspoon vanilla

Filling (follows)

Put the above ingredients in a large bowl.
Work together until the butter is blended.

Filling: 1½ cups chopped Medjools   ½ cup water
½ cup white sugar   2 tablespoons flour
Add dash of lemon juice, cook and cool.

Cover bottom of pan with half of the first mixture. Spread on the filling. Cover with the rest of the mixture. Bake at 350° for approximately 20 minutes in a 6x11-inch or 10x10-inch pan.

Coconut Medjool bars

1 cup chopped Medjool dates	½ cp butter, softened
¾ cup packed brown sugar	1 egg
2/3 cup unsifted flour	1 teaspoon vanilla
½ teaspoon salt	½ tsp baking powder
1 cup rolled oats	1 cup flaked coconut
½ cup chopped nuts	

Preheat oven to 350°. In mixing bowl, cream butter and sugar; beat in egg and vanilla. Mix in remaining ingredients, mixing well. Spread in greased 9-inch square or 6x10-inch pan. Bake 25 to 30 minutes or until toothpick inserted in center comes out clean. Cool and cut into bars. Yield: 36 bars.

Medjool nut kisses

¾ cup chopped Medjool dates   3 egg whites
1 cup sugar   ¼ teaspoon salt
1 teaspoon vanilla   ¾ cup chopped nuts

Combine egg whites, sugar, salt and vanilla in top of double boiler, stir to blend. Place over boiling water and beat until mixture peaks. To prevent being lumpy, scrape bottom and sides of pan occasionally with rubber scraper. Stir in nuts and Medjools at once. Drop heaping teaspoons full of mixture about 2 inches apart onto lightly-greased baking sheets. Bake in slow oven at 300° 12 to 15 minutes or until very lightly browned. Remove from baking sheet immediately and cool on racks.

Medjool-filled cookies II

½ cup shortening   1 cup sugar
2 eggs   1 teaspoon vanilla
2 ½ cups sifted flour   ¼ teaspoon baking soda
                    ½ teaspoon salt
            Chopped Medjool date filling (below)

Mix together shortening, sugar and eggs. Stir in vanilla and mix thoroughly. Sift together flour, soda and salt. Blend into sugar/egg mixture. Chill thoroughly. Roll dough thin, about 1/16", and cut with 2 1/2" round cutter (or use any other shape cutter). Place half of cookies 1" apart on lightly-greased baking sheet. Spread a generous teaspoonful of cooled Medjool filling on each. Cut centers out of other half of cookies, using a small heart, star or other shaped cutter. Place over cookies on baking sheet. Press edges together with floured fork tines or fingers. Bake in oven at 400° 8 to 10 minutes or until lightly browned. Spread on racks to cool.

Medjool date filling: In a small saucepan, combine 2 cups finely-chopped Medjools, ¾ cup sugar and ¾ cup water; cook slowly, stirring constantly until mixture thickens. Remove from heat; stir in ½ cup chopped walnuts (optional) and 1 teaspoon finely grated lemon or orange peel. Cool before serving.

Medjool bars I

½ cup melted butter	1 cup sugar
2 eggs, well beaten	¾ cup flour
¼ teaspoon baking powder	1/8 teaspoon salt
1 cup finely chopped nuts	¼ cup powdered sugar
1 cup finely chopped Medjool dates	

Blend butter, sugar and eggs in bowl. Sift flour with baking powder and salt. Add to egg mixture; mix well. Stir in nuts and Medjools. Spread in greased 8x10-inch baking pan. Bake at 350° for 30 minutes. Cut into 2-inch bars. Sprinkle with confectioner's sugar while warm. Yield: 2 dozen.

Medjool-filled turnovers

½ cup shortening	1 cup sugar
2 eggs	1 teaspoon vanilla
2 ½ cups sifted flour	¼ teaspoon baking soda
½ teaspoon salt	
Chopped Medjool filling below	

Mix together shortening, sugar and eggs. Stir in vanilla and mix thoroughly. Sift together flour, soda and salt. Blend into sugar/egg mixture. Chill thoroughly. Roll dough thin to about 1/16″ and cut with 3″ round cutter. Place one teaspoonful of cooled Medjool filling on each. Fold over and press edges to seal. Place turnovers on baking sheet. Bake in oven at 400° 8 to 10 minutes or until lightly browned. Spread on racks to cool.

Medjool date filling: In a small saucepan, combine 2 cups finely chopped Medjools, ¾ cup sugar and ¾ cup water; cook slowly, stirring constantly until mixture thickens. Remove from heat; stir in ½ cup chopped walnuts (optional) and 1 teaspoon finely-grated lemon or orange peel. Cool before serving.

Medjool nut bars

1 cup chopped Medjool dates	1 cup powdered sugar
1 tablespoon oil	2 eggs, beaten
¼ cup flour	¼ teaspoon salt
½ teaspoon baking powder	¾ cup chopped nuts
1 teaspoon vanilla	Powdered sugar (for tops)

Add 1 cup powdered sugar and oil to eggs; blend well. Add sifted dry ingredients. Stir in nuts, Medjools and vanilla. Pour into greased 9-inch square pan. Bake at 325° for 25 minutes. Cool slightly in pan on rack. Cut in 1x3-inch bars; sprinkle with confectioner's sugar.

Medjool marshmallow goodies

1 ¼ cup chopped Medjools	½ cup butter
1 (10 ½ oz.) pkg tiny marshmallows	2 tablespoons milk
2 cups graham cracker crumbs	1 tablespoon butter
½ cup chopped nuts	1 cup powdered sugar

1 square semisweet dark chocolate

Melt ½ cup butter in 3-qt. Saucepan. Add marshmallows and cook over low heat until melted, stirring constantly. Stir in Medjools, graham cracker crumbs and nuts. Press into buttered 9" square pan. Combine chocolate, milk and 1 tablespoon butter in small saucepan over low heat; stir constantly until chocolate and butter are melted. Stir in powdered sugar. Spread over mixture in pan. Let stand in pan until set, then cut in 1½" squares.

Medjool bars II

½ cups chopped Medjools	4 beaten eggs
1¼ cups sugar	1 cup brown sugar
½ teaspoon salt	2 cups sifted flour
2 teaspoons baking powder	½ teaspoon nutmeg
1 cup chopped almonds	1/3 cp powdered sugar

In a bowl combine flour, baking powder and nutmeg. In another bowl blend eggs, sugar, brown sugar and salt. Gradually mix in flour mixture. Add Medjools and almonds. Spread batter into greased baking pan. Bake at 350° for 30 minutes. Remove pans to a rack. Let stand 5 minutes. Cut into bars. Roll bars in powdered sugar. Let stand to cool. Yield: 5 dozen.

Double Medjool kiddie kookies

2 egg whites	2 cups brown sugar, firmly packed
2 cups sliced Brazil nuts	
1 cup chopped Medjool dates	

Beat egg whites until stiff. Beat in brown sugar gradually. Work in nuts and Medjools. Drop by teaspoonfuls 1" apart onto greased baking sheet. Bake in very slow oven at 250° for 30 minutes. Remove from baking sheet immediately and cool on racks. Yield: 5 dozen.

Medjool and nut roll

2 cups vanilla wafer crumbs (about 8 ounces)
1 cup chopped Medjools
1/2 cup sweetened condensed milk
2 teaspoons lemon juice
1/2 cup chopped nuts

Combine vanilla wafer crumbs, Medjools and chopped nuts. Blend condensed milk and lemon juice; add to crumb mixture and knead well to blend ingredients. Form into rolls about 3 inches in diameter. Roll in more wafer crumbs...then wrap in waxed paper. Chill for at least 12 hours, then cut into 1/4" slices. Serve with whipped cream or dessert sauce for a deliciously different dessert.

Medjool tropical delight

1 cup Medjools pitted & chopped	¼ cup chopped pecans
1 package gelatin	¼ cup cold water
1 cup very strong hot coffee	½ cup sugar
1 tablespoon cocoa	¼ teaspoon salt
½ cup cream (whipped)	½ teaspoon vanilla
1 tablespoon lemon juice	

Pour cold water in bowl and sprinkle gelatin on top of water. Add sugar, salt, cocoa, hot coffee and lemon juice and stir until dissolved. Cool. When it begins to stiffen, add Medjools and nuts. Fold in whipped cream and vanilla. Chill. When firm, unmold and serve with whipped cream. Garnish with stuffed dates. Yield: 6 servings.

Medjool nut squares

1 1/2 cups chopped Medjools	2 1/2 cups sugar
1/4 teaspoon salt	1 cup evaporated milk
2 cups chopped pecans, hazelnuts or walnuts	
1 teaspoon vanilla extract	

In a heavy saucepan, combine sugar, salt and evaporated milk. Bring to a boil, stirring until sugar is dissolved. Cook, stirring occasionally, until mixture reaches soft ball stage or to 236° on candy thermometer. Remove from heat and add remaining ingredients. Let stand until lukewarm, then beat with a wooden spoon until mixture is creamy and loses its gloss. Spread mixture in a 9-inch square pan and let stand until firm. Cut into squares. Yield: 2 pounds of candy.

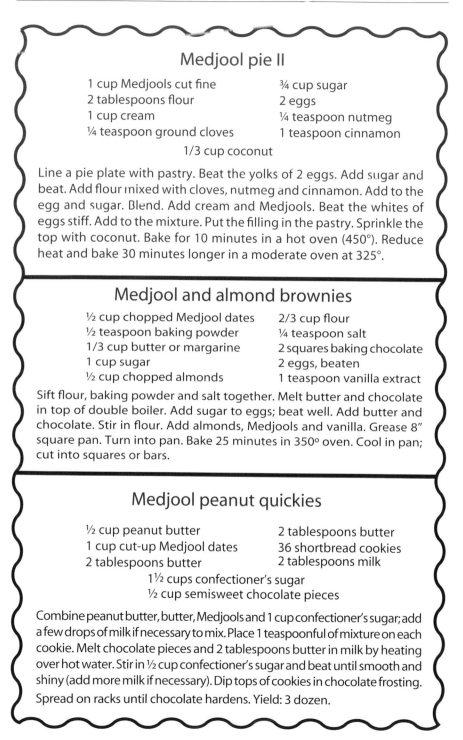

Medjool pie II

1 cup Medjools cut fine	¾ cup sugar
2 tablespoons flour	2 eggs
1 cup cream	¼ teaspoon nutmeg
¼ teaspoon ground cloves	1 teaspoon cinnamon

1/3 cup coconut

Line a pie plate with pastry. Beat the yolks of 2 eggs. Add sugar and beat. Add flour mixed with cloves, nutmeg and cinnamon. Add to the egg and sugar. Blend. Add cream and Medjools. Beat the whites of eggs stiff. Add to the mixture. Put the filling in the pastry. Sprinkle the top with coconut. Bake for 10 minutes in a hot oven (450°). Reduce heat and bake 30 minutes longer in a moderate oven at 325°.

Medjool and almond brownies

½ cup chopped Medjool dates	2/3 cup flour
½ teaspoon baking powder	¼ teaspoon salt
1/3 cup butter or margarine	2 squares baking chocolate
1 cup sugar	2 eggs, beaten
½ cup chopped almonds	1 teaspoon vanilla extract

Sift flour, baking powder and salt together. Melt butter and chocolate in top of double boiler. Add sugar to eggs; beat well. Add butter and chocolate. Stir in flour. Add almonds, Medjools and vanilla. Grease 8" square pan. Turn into pan. Bake 25 minutes in 350° oven. Cool in pan; cut into squares or bars.

Medjool peanut quickies

½ cup peanut butter	2 tablespoons butter
1 cup cut-up Medjool dates	36 shortbread cookies
2 tablespoons butter	2 tablespoons milk

1½ cups confectioner's sugar
½ cup semisweet chocolate pieces

Combine peanut butter, butter, Medjools and 1 cup confectioner's sugar; add a few drops of milk if necessary to mix. Place 1 teaspoonful of mixture on each cookie. Melt chocolate pieces and 2 tablespoons butter in milk by heating over hot water. Stir in ½ cup confectioner's sugar and beat until smooth and shiny (add more milk if necessary). Dip tops of cookies in chocolate frosting. Spread on racks until chocolate hardens. Yield: 3 dozen.

Medjool sticky toffee pudding

This is a dessert with a classic British name. It is actually cake…or, at least that is what most Americans would certainly call it. The Brits seem to have a quaint way of juggling names around to ultimately help them sound more "romantic." Although it seems to have made its way to many locales worldwide, the recipe is attributed by some to an English chef who, according to the story, invented it in the 1960s. There are those who say they will travel long distances to partake of this tasty Medjool magnified (and dignified) treat…and after trying it yourself, we believe you will become a member of that dedicated group.

1 cup finely-chopped Medjool dates	1 cup water
1 stick butter softened	½ cup brown sugar
2 eggs	1 teaspoon vanilla
1 tablespoon golden or corn syrup	1 ¾ cups plus 1 tbsp flour
1 teaspoon baking soda	Whipping cream

1 cup plus 2 tablespoons light cream

Sauce:

1 ½ sticks butter or regular margarine
¼ cup plus 2 tablespoons heavy cream
1 ¼ cups dark brown sugar, firmly packed

Preheat oven to 325°. In a large mixing bowl using an electric mixer, cream butter and brown sugar. Beat in eggs, vanilla and syrup. Stir in flour and dates.

Dissolve baking soda in light cream and then add to batter, blending well.

Grease a 7x11 or 8x8x2" pan. Pour in batter and bake about 40 to 50 minutes or until the top turns golden brown and cracks slightly, the sides pull away slightly from the edge of the pan, the cake feels springy when touched and a toothpick inserted in the center comes out clean.

While the cake is baking, make the sauce. Place topping ingredients in a heavy saucepan and bring to a boil. Simmer 3 to 5 minutes or until thickened.

Remove pan from heat. Pour a generous ¼ cup of the sauce over the top of the cake and place the cake under the broiler till the topping is lightly browned and bubbly.

Invert cake onto a serving plate and cut into pieces. Spoon some warm sauce onto individual plates and put slices of warm cake on top. Serve with a dollop of freshly-whipped cream. Serves: 10

Medjool chocolate nut bread

1 cup sliced pitted Medjool dates	¾ cup boiling water
1 egg	¾ cup milk
1 teaspoon vanilla extract	2 ½ cups flour
1/3 cup sugar	1 ½ teaspoons salt
1 teaspoon baking powder	1 teaspoon baking soda

4-ounce can walnuts (1 cup), coarsely chopped
1 6 oz. package semi-sweet chocolate pieces (1 cup)
4 tablespoons butter or margarine (½ stick)

Preheat oven to 350°. Grease 9"x5" loaf pan. In small bowl, pour boiling water over Medjools; set aside. In heavy 1-qt. saucepan over low heat, melt chocolate with butter or margarine. In small bowl, beat egg with milk and vanilla just until mixed.

In large bowl, mix flour and remaining ingredients; add Medjools and liquid, chocolate and milk mixtures; mix just until blended. Spoon batter evenly into loaf pan.

Bake 1 hour and 10 minutes or until toothpick inserted in center comes out clean. Cool bread in pan on wire rack 10 minutes; remove from pan and finish cooling on wire rack.

Medjool orange salad I

1 cup Medjool dates chopped	1 cup orange juice
1 package orange gelatin	1 cup hot water
½ cup pecan halves	1 fresh avocado
4-6 fresh oranges, peeled and cut up	Mayonnaise

Dissolve gelatin in hot water; add orange juice. Pour into a flat glass container; add orange pieces, Medjools and nuts. Mix well. Chill until firm. Mash avocado with enough mayonnaise to make a creamy mixture; season to taste. Spoon into mound over salad.

Medjool orange salad II

1 cup Medjool dates chopped	1 package orange gelatin
1 8-oz. package cream cheese	1 cup hot water
½ cup orange juice	2 teaspoons lemon juice

1 cup orange sections

Dissolve gelatin in hot water; gradually add cream cheese. Blend until smooth. Add orange and lemon juice. Chill until slightly thickened. Add orange sections and Medjools. Chill until firm.

Bacon-wrapped Medjools stuffed with almonds

1 pound pitted Medjools
1 packet blanched whole almonds
1 ½ pound lean thinly-sliced bacon cut into thirds (4-oz.)

Insert an almond into each Medjool. Wrap bacon around each stuffed date and secure with a toothpick. Place dates on cookie sheet and bake in preheated 400 degree oven for 12-15 minutes or until bacon is crisp. Check frequently for condition. Remove to rack or paper towel; drain. Serve warm.

Medjool skillet cookies

1 cup chopped Medjool dates	1 cup sugar
3 tablespoons butter	1 egg, well beaten
1/2 teaspoon vanilla	¾ cup pecans
1 cup flaked coconut	2 cups crisp rice cereal

Mix Medjools, sugar, butter and egg in heavy skillet. Melt over low heat; cook for about 5 minutes on low heat, until bubbly. Remove from heat; add vanilla, rice cereal and pecans. When cool enough to handle, roll into small balls and roll in coconut.

Medjool pudding II

1 cup pitted Medjool dates	1 cup walnut meats
1 cup powdered sugar	2 eggs
1 teaspoon baking powder	Vanilla
5 tablespoons flour	

Put the Medjools and walnuts through a food processor. Add the sugar and the eggs (well beaten.) Mix, add flour and baking powder. Flavor with vanilla. Spread in a large shallow pan, ¼" thick. Bake in a moderate oven at 325°. While hot, cut into narrow strips. Roll in powdered sugar. These may be eaten plain or, if desired for pudding, they may be broken into pieces and mixed with whipped cream. Serve in sherbet glasses. Serves 6.

Medjool orange cake

1 lb. chopped Medjools	1 1/2 cups butter
6 eggs, separated	3 cups granulated sugar
2 cups buttermilk	3 tbsp. grated orange peel
1 1/2 tsp. baking soda	1/2 tsp salt
2 1/2 cups pecans, chopped	6 cups flour

Cream butter and 1 1/2 cups sugar until light. Beat egg yolks until lemon colored and add to creamed mixture. Add buttermilk and grated orange peel; mix well. Mix dates and pecans with 1 cup of the flour. Mix remaining flour with other dry ingredients and add to creamed mixture along with nuts and dates. Beat egg whites until stiff; fold into batter. Pour into greased and floured pan and bake at 400° for one hour and fifteen minutes or until toothpick tester inserted in middle of cake comes out clean. Remove from oven and glaze while hot with a mixture of 3 cups sugar, orange juice and 3 tablespoons grated orange peel.

Glaze:
3 cups granulated sugar
3/4 cup orange juice

Medjool nut bread II

8 ounces chopped Medjools	1 cup boiling water
2 tablespoons butter	¾ cup granulated sugar
¼ cup brown sugar	1 large egg
2 ¼ cups flour, stir before measuring	½ teaspoon salt
1 tablespoon baking powder	½ cup chopped walnuts

Pour boiling water over Medjools in a medium size bowl; add butter and set aside. With mixer, beat sugars and egg until light. In another bowl, combine the flour, baking powder and salt. Add to the sugar mixture, alternating with the Medjool and water mixture. Stir in chopped nuts. Grease a 9x5x3" loaf pan and fill. Bake at 325° for 45 to 55 minutes or until wooden pick inserted in center comes out clean.

Moroccan Medjool morsels

¾ cup Medjools pitted, chopped   2 packages gelatin
¼ cup raisins, chopped   ½ cup chopped nuts
½ cup cold water   2 cups sugar
¾ cup boiling water   2 tblespoons lemon juice
¼ teaspoon salt

Heat sugar, salt and water to boiling point. Pour cold water in bowl and sprinkle gelatin on top of water. Add to hot syrup and stir until gelatin is dissolved. Boil slowly for 15 minutes. Add nuts, raisins, Medjools and lemon juice. Pour mixture into pan (about 4x8") that has been rinsed in cold water. Cool for at least 12 hours in cold place (not refrigerator) until thick and firm. With knife, loosen edges of pan, turn out on a board lightly covered with powdered sugar. Cut in cubes and roll in powdered sugar.

Medjool pumpkin cookies

1 cup chopped Medjool dates   1 cup rolled oats
½ cup chopped nuts   1 ½ cups whole-wheat flour
2 teaspoons baking powder   ½ teaspoon baking soda
½ teaspoon salt   1 teaspoon cinnamon
½ teaspoon nutmeg   1/8 teaspoon ground cloves
2/3 cup butter   ½ cup sugar
½ cup packed brown sugar   2 eggs, beaten
1 ¼ cups canned pumpkin   1 teaspoon vanilla

Mix flour, baking powder, soda, salt, nutmeg, cinnamon and cloves until well blended. Cream butter and sugars. Add eggs; beat well. Stir in pumpkin and vanilla. Blend in flour mixture; stir thoroughly. Add Medjools, oats and nuts; mix well. Drop by teaspoons onto greased baking sheet. Bake at 375° about 12 minutes. Remove to rack to cool.

Medjool filled oatmeal cookies

1 c. brown sugar - 1 c. butter - 3 eggs - 2 c. oatmeal
2 1/2 c. flour - 3/4 c. sour milk - 1 tsp. soda in flour
Filling:
1 pkg. Medjools, finely chopped   1 c. brown sugar
1 c. water

Cook filling until thickened. Cream butter and sugar. Add eggs and mix well. Add flour and soda with sour milk...add oatmeal. Roll out on floured board, cut with cookie cutter. Put one teaspoon of filling on cookie and cover with another cookie. Bake at about 375 degrees for 10 to 15 minutes.

Medjool banana bread

1 cup finely chopped Medjools   1 3/4 cups sifted flour
2 3/4 teaspoons baking powder   1/2 teaspoon salt
1/3 cup shortening   2 eggs, slightly beaten
1 cup mashed bananas

Blend together flour, baking powder and salt. Beat the shortening until creamy. Beat in sugar and continue beating until light and fluffy. Add eggs and beat until thick and pale lemon in color. Add Medjools. Add flour mixture and bananas alternately, blending well after each addition. Pour batter into greased loaf pan. Bake at 350° for 60 minutes or until a wooden pick inserted in center comes out clean. Cool in pan for 20 minutes then turn out onto a rack to cool completely.

Stuffed Medjools I

Carefully pit Medjool dates by pushing pit out one end with blunt end of a bamboo spear. Soak pitted Medjool dates in: wine, rum, whiskey, lemon or orange juice. Fill them with pieces of marshmallow or nut meats, candied ginger or candied fruits. Shape the Medjools into their original form and roll them in powdered sugar.

Medjool nut drops

2 cups chopped Medjool dates   ½ cup sugar
1 cup butter or margarine   ½ cup water
1 cup brown sugar, firmly packed   1 cup sugar
3 eggs   1 teaspoon vanilla
4 cups sifted flour   1 teaspoon baking soda
1 teaspoon ground cinnamon   1 teaspoon salt
1 ½ cups chopped walnuts

Combine Medjools, ½ cup sugar and water in saucepan. Cook, stirring occasionally until mixture is the consistency of very thick jam. Cool. Cream butter; add sugars gradually, beating until light and fluffy. Beat in eggs and vanilla. Sift together dry ingredients. Add to creamed mixture, blending thoroughly. Stir in nuts and Medjool mixture. Drop by rounded teaspoonfuls about 2" apart onto greased baking sheet. Bake in oven at 375° for 12 to 15 minutes. Remove cookies and cool on racks. Yield: 6 dozen.

Medjool crumbles

1 cup Medjools chopped small	2 eggs, well beaten
2 teaspoons baking powder	1 cup sugar
1 tablespoon flour	1 cup chopped walnuts

Mix all together and spread on two greased pie tins. Bake in a slow oven three-quarters of an hour. Crumble and serve in tall glasses topped with whipped cream or mix with whipped cream and serve.

Medjool pudding III

20 Medjool dates pitted and chopped	4 egg whites (beaten stiff)
1 teaspoon lemon juice	½ cup powdered sugar

Cover Medjools with water and cook until very soft then mash through a strainer. Fold sugar and Medjool pulp into beaten whites of eggs and add lemon juice. Pile lightly in a buttered pudding dish, cover and set in a pan of hot water in a very moderate oven for 10 minutes then remove cover and cook until top is a light brown. Serve with whipped cream or a soft custard.

Medjool nut strips

1 cup Medjool dates chopped	1 cup chopped walnuts
3 eggs	2 tablespoons water
½ teaspoon salt	½ teaspoon cinnamon
1 teaspoon baking powder	1 cup sugar

½ teaspoon maple flavoring

Beat eggs until light then add 2 tablespoons warm water and beat until thick and lemon-colored. Add the sugar, then fold in flour, baking powder, salt and cinnamon mixed and sifted together. Last…add maple flavoring, chopped walnut meats and Medjools. Spread on a greased pan and bake in moderate oven. Cut into strips while hot; sprinkle with sugar and serve.

Medjool malt

1 cup cold milk	½ pint vanilla ice cream
1 tablespoon malted milk powder	
¼ cup chopped Medjool dates	

Combine all ingredients in an electric blender or shake mixer. Blend or mix at high speed for 1 minute or until smooth and fluffy. Serves: 1

Medjool carrot cake

3/4 cup chopped Medjools
1 1/2 cups vegetable oil
2 cups grated carrots
2 teaspoons baking powder
2 teaspoons ground cinnamon
1/4 teaspoon ground nutmeg
1/2 teaspoon salt

4 eggs
2 cups sugar
2 teaspoons vanilla
3 cups flour
2 teaspoons baking soda
1/8 tsp ground cloves
3/4 cup chopped nuts

Beat together eggs, sugar, oil, grated carrots and vanilla. Sift together the flour, baking soda, baking powder, salt and spices; gradually add to carrot mixture and beat to blend well. Fold in chopped nuts and Medjools. Bake in a greased 10-inch tube pan at 375° for about 55 to 65 minutes. A wooden pick inserted in center should come out clean.
Frost with cream cheese frosting.

Orange Medjool oatmeal drops

¼ cup finely-chopped Medjools
1 cup brown sugar
1 cup sifted flour
2 tablespoons orange juice
½ teaspoon salt
1 cup rolled oats

½ cup butter
1 egg
½ teaspoon baking soda
1 tbsp grated orange rind
½ cup raisins

Combine butter and sugar until creamy. Beat in egg. In another bowl, combine flour, baking soda and salt. Add orange juice, rind, oats and raisins. Blend into egg mixture. Add Medjools. Using a teaspoon, form dough into separate cookies on greased cookie sheets. Bake at 350° for 12 minutes or until lightly-browned. Remove sheets to a rack. Let stand

Mini Medjool drops

1 cup chopped Medjools   2 egg whites
2 cups brown sugar, firmly-packed
2 cups sliced Brazil nuts

Beat egg whites until stiff. Beat in brown sugar gradually. Work in nuts and Medjools. Drop by teaspoonfuls 1" apart onto greased baking sheet. Bake in very slow oven at 250° for 30 minutes. Remove from baking sheet immediately and cool on racks. Yield: 5 dozen.

Medjool half moons

1 (3 oz.) pkg cream cheese	1 cup cookie dough
2 tablespoons powdered sugar	½ teaspoon vanilla
24 pitted Medjool dates	Sifted powdered sugar

Combine first 4 ingredients. Form into four balls. Chill dough 2 hours. Roll balls flat 1/8" thick. Cut in rounds with 2 1/2 " cutter. Place Medjool date in center of each round. Fold edges over and pinch ends to points. Place 1" apart on lightly greased baking sheet, seam side down. Bake in moderate oven at 350° 10 to 12 minutes. Sprinkle with confectioner's sugar. Remove cookies and cool on racks. Yield: 2 dozen.

Medjool cake

½ lb. chopped Medjools (1 ½ cups)	1 cup boiling water
½ cup shortening	1 cup sugar
1 teaspoon vanilla	1 egg
1 ½ cups sifted all-purpose flour	1 teaspoon soda
¼ teaspoon salt	½ cup chopped walnuts

Combine Medjools with water; cool to room temperature. Cream shortening and sugar until light. Add vanilla and egg; beat well. Sift flour, soda and salt together; add to creamed mixture alternately with Medjool mixture, beating after each addition. Stir in nuts. Bake in greased and lightly-floured 13x9x2" baking pan at 350° about 25 to 30 minutes. If desired, serve with a dollop of whipped cream.

Spiced Medjool bars

1 cup chopped Medjools	¼ cup shortening
½ cup sugar	1 egg
2 cups sifted all-purpose flour	½ cup molasses
¼ teaspoon salt	¼ teaspoon soda
1 ½ teaspoons baking powder	1½ tsps grnd cinnamon
¼ teaspoon ground cloves	1/8 teaspoon grnd ginger
1 cup chopped walnuts/pecans	½ cup milk

Cream shortening with sugar; beat in eggs and molasses. Sift together the flour, salt, soda, baking powder and spices. Stir sifted dry ingredients into the creamed mixture alternating with the milk. Stir in nuts and Medjools. Spread mixture in a greased and floured 8x12" baking pan. Bake at 350° for 25 to 30 minutes. Cool in pan and cut into bars. Optional: spread bars with vanilla icing before cutting.

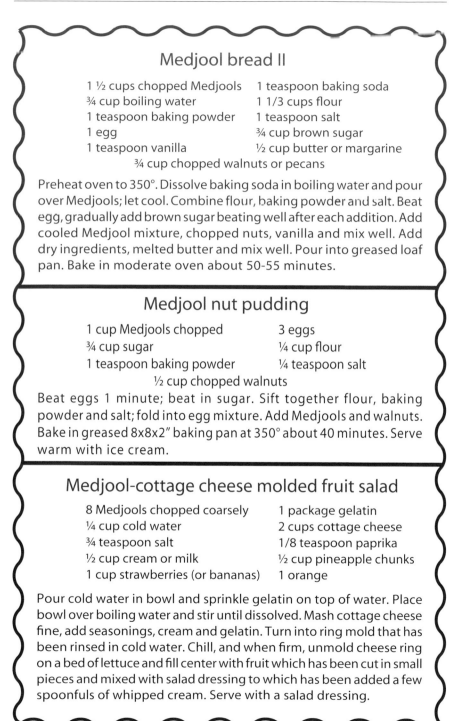

Medjool bread II

1 ½ cups chopped Medjools	1 teaspoon baking soda
¾ cup boiling water	1 1/3 cups flour
1 teaspoon baking powder	1 teaspoon salt
1 egg	¾ cup brown sugar
1 teaspoon vanilla	½ cup butter or margarine

¾ cup chopped walnuts or pecans

Preheat oven to 350°. Dissolve baking soda in boiling water and pour over Medjools; let cool. Combine flour, baking powder and salt. Beat egg, gradually add brown sugar beating well after each addition. Add cooled Medjool mixture, chopped nuts, vanilla and mix well. Add dry ingredients, melted butter and mix well. Pour into greased loaf pan. Bake in moderate oven about 50-55 minutes.

Medjool nut pudding

1 cup Medjools chopped	3 eggs
¾ cup sugar	¼ cup flour
1 teaspoon baking powder	¼ teaspoon salt

½ cup chopped walnuts

Beat eggs 1 minute; beat in sugar. Sift together flour, baking powder and salt; fold into egg mixture. Add Medjools and walnuts. Bake in greased 8x8x2" baking pan at 350° about 40 minutes. Serve warm with ice cream.

Medjool-cottage cheese molded fruit salad

8 Medjools chopped coarsely	1 package gelatin
¼ cup cold water	2 cups cottage cheese
¾ teaspoon salt	1/8 teaspoon paprika
½ cup cream or milk	½ cup pineapple chunks
1 cup strawberries (or bananas)	1 orange

Pour cold water in bowl and sprinkle gelatin on top of water. Place bowl over boiling water and stir until dissolved. Mash cottage cheese fine, add seasonings, cream and gelatin. Turn into ring mold that has been rinsed in cold water. Chill, and when firm, unmold cheese ring on a bed of lettuce and fill center with fruit which has been cut in small pieces and mixed with salad dressing to which has been added a few spoonfuls of whipped cream. Serve with a salad dressing.

Medjool nut Scotch cake

1 cup chopped Medjools	1 cup quick-cooking rolled oats
1 cup boiling water	1 cup unsifted flour
1½ cups brown sugar	1 teaspoon salt
1 teaspoon soda	1 teaspoon cream of tartar
1 teaspoon cinnamon	½ teaspoon ground cloves
½ cup chopped nuts	½ cup shortening or margarine
2 eggs	

Preheat oven to 350°. In large mixing bowl, pour boiling water over rolled oats; stir to moisten. Cool. Add remaining ingredients except dates and nuts to cooled rolled oats mixture. Blend at low speed until moistened; beat 3 minutes at medium speed, scraping bowl occasionally. Stir in dates and nuts. Pour batter into 9-inch square or 7x11-inch pan greased on bottom only. Bake 30 to 35 minutes or until top springs back when lightly touched in center. Serve warm or cool, plain or with whipped cream.

Medjool muffins II

1 cup chopped Medjool dates	1/3 cup butter or margarine
1 egg	2 cups flour
2 teaspoons baking powder	¾ cup milk

1/2 teaspoon salt

Cream butter, add beaten egg, flour in which baking powder and salt have been sifted and milk. Stir in Medjools. Bake about 25 minutes in greased muffin pans in hot oven. For sweet muffins, sift ¼ cup sugar with dry ingredients.

Medjool pecan waffles

1 cup finely chopped Medjool dates	¼ cup chopped pecans
2 cups flour	¼ cup brown sugar
1 tablespoon baking powder	1 teaspoon salt
1 ¾ cups milk	2 eggs, separated

½ cup butter or margarine, melted

Combine flour, sugar, pecans, baking powder and salt in large mixing bowl. Add the Medjools. Combine the milk, egg yolks and butter in a small bowl. Stir them into the dry ingredients.

In a small mixing bowl, beat egg whites into stiff peaks, then fold into batter. Bake in a preheated waffle iron until golden brown.

Topping: 1 pkg (8 oz.) cream cheese, softened
     ¼ cup ½ & ½ cream   2 tablespoons sugar
   2 tablespoons orange juice   1 tablespoon grated orange peel

In a mixing bowl, beat the ingredients until blended. Serve with waffles.

Medjool roll wafers

1 cup shortening
1 cup granulated sugar
4 cups all-purpose flour
1/2 teaspoon salt

1 cup brown sugar, packed
3 large eggs
1 teaspoon baking soda
1 teaspoon vanilla

Filling:

8 ounces chopped Medjools   3/4 cup granulated sugar
1/2 cup boiling water

Cream shortening and add sugar. Beat in eggs; add dry ingredients and vanilla until dough is formed. Chill dough thoroughly. Combine filling ingredients in a saucepan; boil until thick. Let cool to room temperature. Roll out the chilled cookie dough into a rectangle between two sheets of waxed paper. Spread the cooled date filling over the cookie dough. Roll up the cookie dough, jellyroll fashion. Wrap in waxed paper and chill before slicing. Slice and bake at 375°.

Yuma Medjool pie

1 pound Medjool dates
½ cup sugar
1 egg, well beaten

1 cup milk
1 tablespoon flour
1 baked pastry shell, 9"

Whipped cream topping

Pit and chop Medjools coarsely. Place in saucepan, cover with water and simmer covered until soft. Mix together sugar, milk, dash of salt and flour; add to date mixture then add beaten egg. Cook until thickened, stirring constantly. Cool and pour into pastry shell. When firm, cover Medjool pie with whipped cream topping.

Medjool cornflake drop cookies

2/3 cup Medjools, pitted, chopped
1 2/3 cups crushed cornflakes
2/3 cup chopped walnuts or pecans
½ teaspoon vanilla

2 eggs
1/3 cup sugar
A pinch of salt

Beat the 2 eggs, then add gradually the 1/3 cup sugar. When blended, stir in the remaining ingredients and let stand for 30 minutes. Drop by spoonful on greased cookie sheet. Bake in a moderate oven at 350° for about 10 minutes.

Medjool coconut cookies

1/2 lb. chopped Medjools   1 cup chopped pecans
1 can (3.5 ounces) flaked coconut 1 egg
          1/2 cup firmly-packed brown sugar

Add together dates and pecans; add half of the coconut, brown sugar and egg. Mix well. Shape into 2-inch rounds then roll them in remaining coconut. Place date cookies on greased baking sheets and bake at 350° for 10 to 15 minutes.

Medjool double nut drops

½ pound chopped Medjool dates 1 cup white sugar
3 beaten egg whites   ¼ pound pecans
1 teaspoon vanilla extract   ¼ pound walnuts

Blend beaten egg whites and sugar in double-boiler top. Cook over boiling water until mixture is stiff. Remove from heat. Combine with pecans, dates and vanilla. Using a teaspoon, form drops about 2-inches apart on greased cookie sheets. Bake at 320° for 45 minutes.

Medjool Christmas pudding

2/3 cup Medjool dates chopped   ½ cup raisins
¼ cup currant jelly   ¼ cup nuts, chopped
1 package gelatin   ½ cup cold water
1 cup milk   3 tablespoons cocoa
½ cup sugar   ¼ teaspoon salt
¼ teaspoon vanilla   2 egg whites

Put milk with chopped Medjools in double boiler. When cooked slightly, add cocoa which has been mixed with part of the sugar and a little milk to make a smooth paste. Pour cold water in bowl and sprinkle gelatin on top of water. Add to hot chocolate mixture and stir until dissolved. Add sugar and salt and stir thoroughly. Remove from heat and cool...and when mixture begins to thicken, add nuts and vanilla. Lastly, fold in whites of eggs beaten very stiff. Turn into mold that has been rinsed in cold water and decorated with whole nut meats and raisins placed at bottom of mold. Chill. When firm, unmold on serving dish and garnish with holly. Serve with whipped cream sweetened with vanilla or with a currant jelly sauce. Serves 6.

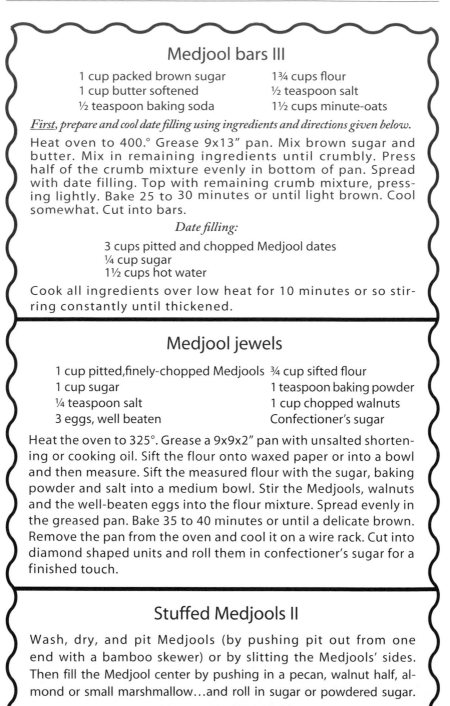

Medjool bars III

1 cup packed brown sugar	1¾ cups flour
1 cup butter softened	½ teaspoon salt
½ teaspoon baking soda	1½ cups minute-oats

First, prepare and cool date filling using ingredients and directions given below.

Heat oven to 400.° Grease 9x13" pan. Mix brown sugar and butter. Mix in remaining ingredients until crumbly. Press half of the crumb mixture evenly in bottom of pan. Spread with date filling. Top with remaining crumb mixture, pressing lightly. Bake 25 to 30 minutes or until light brown. Cool somewhat. Cut into bars.

Date filling:

3 cups pitted and chopped Medjool dates
¼ cup sugar
1½ cups hot water

Cook all ingredients over low heat for 10 minutes or so stirring constantly until thickened.

Medjool jewels

1 cup pitted,finely-chopped Medjools	¾ cup sifted flour
1 cup sugar	1 teaspoon baking powder
¼ teaspoon salt	1 cup chopped walnuts
3 eggs, well beaten	Confectioner's sugar

Heat the oven to 325°. Grease a 9x9x2" pan with unsalted shortening or cooking oil. Sift the flour onto waxed paper or into a bowl and then measure. Sift the measured flour with the sugar, baking powder and salt into a medium bowl. Stir the Medjools, walnuts and the well-beaten eggs into the flour mixture. Spread evenly in the greased pan. Bake 35 to 40 minutes or until a delicate brown. Remove the pan from the oven and cool it on a wire rack. Cut into diamond shaped units and roll them in confectioner's sugar for a finished touch.

Stuffed Medjools II

Wash, dry, and pit Medjools (by pushing pit out from one end with a bamboo skewer) or by slitting the Medjools' sides. Then fill the Medjool center by pushing in a pecan, walnut half, almond or small marshmallow…and roll in sugar or powdered sugar.

Steamed Medjool pineapple pudding

1 cup chopped Medjool dates
½ cup sugar
1 cup pineapple juice
½ cup chopped nuts
1½ cups flour
¾ teaspoon baking soda
3 tablespoons melted butter or margarine
1 teaspoon cinnamon

1 egg
½ teaspoon salt
¼ cup chopped raisins
1 teaspoon vanilla
1 teaspoon baking powder
¼ teaspoon nutmeg

Beat the egg well, add the sugar, salt, butter and pineapple juice. Heat. Then stir in the raisins, nuts, Medjools and vanilla. Mix well. Sift together the dry ingredients and add to the mixture. Pour into a well-greased mold, cover tightly and steam for 2 hours. Unmold and serve.

Helen's Orange County cake bread

1/2 cup Medjools, chopped
1 egg
2 cups flour
3/4 tsp salt
1/4 cup water
1 teaspoon baking powder

1 cup sugar
2 tbsp butter, melted
1/4 tsp baking soda
1/2 cup orange juice
1 cup pecans, chopped

Soak Medjools for 2 hours in water, drain and dry on paper toweling. In a bowl, mix egg, sugar and butter. Sift flour with baking powder, baking soda and salt. Add dry ingredients alternately with orange juice and water beginning and ending with dry ingredients. Stir in Medjools and pecans and beat until well blended. Pour mixture into a greased 9x5x3" loaf pan. Bake in a preheated 350° F. oven for about 2 hours or until bread tests done in the center.

Medjool holiday cups

2 cups chopped Medjool dates
1 ½ teaspoons soda
1 egg
¼ teaspoon salt
1 teaspoon vanilla extract

1 cup chopped pecans
¾ cup butter
1 ½ cup flour
1 cup whipping cream
1 cup sugar

Mix Medjools, pecans, 1 teaspoon soda and 1 cup boiling water in bowl. Let stand for 10 to 15 minutes. Cream ¼ cup butter and 1 cup sugar in mixer bowl until fluffy. Mix in egg. Sift in flour, salt and remaining ½ teaspoon soda; mix well. Stir in Medjool mixture. Fill 18 muffin cups 2/3 full. Bake at 350° for 35 minutes. Cook remaining ½ cup butter, 1 cup sugar, cream and vanilla in saucepan for 10 minutes, stirring constantly. Serve over warm date cups.

Medjool Candy

1 cube margarine
1 beaten egg
3 cups Rice Krispies
1/2 c. chopped walnuts

1 cup chopped Medjools
1 cup packed brown sugar
1 teaspoon vanilla
1 cup of coconut

Mix margarine, brown sugar, Medjools and egg in heavy pan; cook slowly until thick and stir often. Remove from heat then add vanilla, nuts and Rice Krispies. Let cool and roll into balls. Then roll in shredded coconut. Place on wax paper and chill.

Compliments of Vicky Daniels

Medjool tropical pound cake

1/2 cup Medjools, chopped
1 cup butter
1 1/2 cups flour
2 tbsps grated orange peel

1 1/2 cups sugar
4 eggs
1 1/2 tsps baking powder
2 tbsps orange juice

1/2 cup chopped pecans

In a bowl, beat butter, then add sugar gradually...beating until light and fluffy. Add eggs one at a time, beating after each. In a separate bowl, stir together flour and baking powder. Add orange peel and orange juice; fold in pecans and Medjools. Grease a 9x5x3-inch loaf pan; dust with flour. Pour batter into loaf pan. Bake at 350° for 45 to 55 minutes or until cake tests done. Cool on rack.

Medjool Fridge Cookies

1 cup chopped Medjools
1/2 cup butter
1 teaspoon vanilla
3 1/2 cups flour
1 cup chopped walnuts

1/2 cup shortening
2 cups light brown sugar
2 large eggs
3/4 teaspoon salt

Cream shortening, butter and sugar; beat in vanilla and egg until light and fluffy. Sift flour and salt; add to creamed mixture. Stir in Medjools and walnuts. Shape into two 2-inch rolls; wrap in waxed paper and chill overnight. Slice cookie dough into thin slices and bake on greased baking sheets at 375° for 8 to 10 minutes. Yield: 6 dozen date nut cookies.

Medjool oasis bits

3/4 cup margarine	1 3/4 cups all-purpose flour
1 teaspoon salt	1/2 teaspoon soda
1 cup light brown sugar, firmly packed	
1 1/2 cups quick-cooking oats, uncooked	

Cream margarine and sugar together until light and fluffy. Combine flour, salt and soda; add to creamed mixture. Stir in oats, blending well. Pat half of the mixture into a lightly buttered 13x9x2-inch baking pan; spread over with Medjool filling, then top that with remaining oat mixture. Smooth top crust. Bake at 400° for 30 minutes. Cut into small units while still warm; remove from pan.

Medjool filling:

1 1/4 pounds Medjools, chopped, 1/4 cup sugar,1 1/2 cups water. Combine dates, sugar and water in a medium saucepan; cook over low heat for 5 to 10 minutes until smooth and thick. Cool slightly, then spread on 1st layer of oat mix. Makes 2 cups of date filling for date bits.

Medjool walnut granola bars

Filling:

1 ½ cups chopped Medjools	¾ cup half and half
½ cup shredded coconut	1 teaspoon vanilla

Granola crust:

½ cup chopped walnuts	¾ cup all-purpose flour
2 cups lightly toasted quick oats	½ c. melted salted butter
1 cup dark brown sugar	½ teaspoon baking soda
1 teaspoon ground cinnamon	

Filling:

Place dates, coconut and half and half in saucepan over medium heat. Stir until mix boils and thickens…about 15 minutes and remove from heat. Stir in vanilla and set aside to cool.

Crust:

Mix together oats, flour, sugar, soda and cinnamon in a medium bowl. Blend well. Pour melted butter over all and stir until well moistened. Press about 3 cups granola mixture into bottom of 8 inch square baking pan and chill 30 minutes to harden. To bake: Preheat oven to 350°. Spread cooled filling over granola base. Sprinkle remaining granola mixture over filling. Sprinkle finely-chopped walnuts over all. Bake 30 minutes or until top is slightlybrowned and crisp. Cool to room temperature and cut into bars.

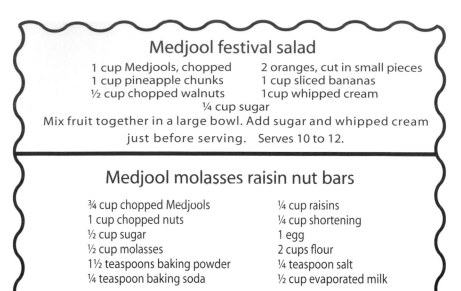

Medjool festival salad

1 cup Medjools, chopped	2 oranges, cut in small pieces
1 cup pineapple chunks	1 cup sliced bananas
½ cup chopped walnuts	1 cup whipped cream

¼ cup sugar

Mix fruit together in a large bowl. Add sugar and whipped cream just before serving.   Serves 10 to 12.

Medjool molasses raisin nut bars

¾ cup chopped Medjools	¼ cup raisins
1 cup chopped nuts	¼ cup shortening
½ cup sugar	1 egg
½ cup molasses	2 cups flour
1½ teaspoons baking powder	¼ teaspoon salt
¼ teaspoon baking soda	½ cup evaporated milk

Cream shortening, add sugar and beat until light. Add beaten egg, mix well, then add molasses. Mix flour with dry ingredients and add with milk to first mixture. Add chopped nuts, Medjools and raisins last. Spread thinly in shallow pan. Bake in moderate oven (350º) from 15 to 20 minutes. Cut in bars 2 ½ inches long and 1 ½ inches wide before removing from pan. Makes about 4 dozen.

Medjool tarts

1 cup finely chopped Medjools	3 cups flour
1¼ cups butter, melted-divided	2 eggs, beaten
¾ cup ground blanched almonds	¼ cup sugar
½ cup milk	½ teaspoon salt
1½ teaspoons vanilla	2 tsps baking powder
½ cup cinnamon	1¼ teaspoon nutmeg
1/2 teaspoon ground cloves	½ cup powdered sugar

Combine 1 cup butter, eggs, milk, salt and vanilla. Sift flour and baking powder in gradually. Knead until soft dough is formed adding water as necessary. Form into 1" balls, cover with a cloth and allow to rest for 1 hour.

Filling: Thoroughly mix the remaining ingredients including the remaining ¼ cup of butter.

Flatten the balls to a 3" or 4" diameter, then place a heaping teaspoon of filling on each center. Fold over to cover the filling and pinch edges together to form tart. Place on baking tray; bake for 20 minutes at 350º or until tarts turn golden brown. Remove from heat and cool. Sprinkle with powdered sugar and serve.

Medjool sandwich spreads

1. Mix finely-chopped Medjools with cream cheese.

2. Mix equal parts of finely-chopped Medjools, peanut butter and mayonnaise. Mix all with modest amount of lemon juice.

3. Mix finely-chopped Medjools with orange marmalade.

Medjool party snacks

1. Pit Medjools, stuff with olives rolled in thin cream cheese and finely chopped pecans.

2. Stuff pitted Medjools with fondant, nutmeats or candied fruit.

3. Medjool "Chews." Mix finely chopped Medjools, finely chopped nuts, shredded coconut, cream and vanilla. Make into balls and coat with sugar.

Medjool filled cookies

½ cup cut-up Medjool dates
1 cup orange juice (or water)
1 teaspoon grated orange peel
2 tablespoons sugar
30 to 40 small cookies, preferably "oatmeal"

Blend Medjools, sugar, orange juice and orange peel in small saucepan. Cook over medium heat stirring constantly until thick. Place 1 scant teaspoon of Medjool filling between pairs of oatmeal cookies.

Stuffed Medjools III

1 pound Medjools	½ cup walnut halves
¾ cup half & half cream	1 cup sugar
4 tbsps cocoa	1 cup shredded coconut

1 tbsp mazahar (orange flower water - found at Mideast markets)

Slit the Medjools on a side, then stuff with the walnut halves. Set aside. Place cream in a small pot, then bring it to a boil over medium heat. Add the sugar and stir until it melts. Add mazahar and cocoa stirring constantly; cook for about 5 minutes. Remove and allow to cool. Dip Medjools in cocoa syrup, then roll in coconut.

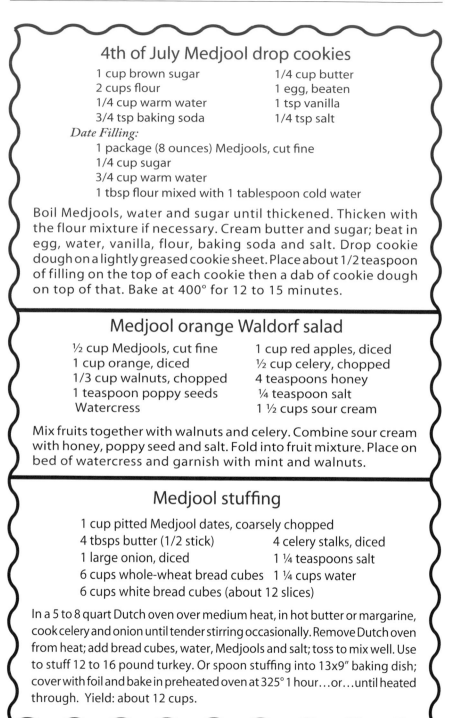

4th of July Medjool drop cookies

1 cup brown sugar	1/4 cup butter
2 cups flour	1 egg, beaten
1/4 cup warm water	1 tsp vanilla
3/4 tsp baking soda	1/4 tsp salt

Date Filling:

1 package (8 ounces) Medjools, cut fine
1/4 cup sugar
3/4 cup warm water
1 tbsp flour mixed with 1 tablespoon cold water

Boil Medjools, water and sugar until thickened. Thicken with the flour mixture if necessary. Cream butter and sugar; beat in egg, water, vanilla, flour, baking soda and salt. Drop cookie dough on a lightly greased cookie sheet. Place about 1/2 teaspoon of filling on the top of each cookie then a dab of cookie dough on top of that. Bake at 400° for 12 to 15 minutes.

Medjool orange Waldorf salad

½ cup Medjools, cut fine	1 cup red apples, diced
1 cup orange, diced	½ cup celery, chopped
1/3 cup walnuts, chopped	4 teaspoons honey
1 teaspoon poppy seeds	¼ teaspoon salt
Watercress	1 ½ cups sour cream

Mix fruits together with walnuts and celery. Combine sour cream with honey, poppy seed and salt. Fold into fruit mixture. Place on bed of watercress and garnish with mint and walnuts.

Medjool stuffing

1 cup pitted Medjool dates, coarsely chopped

4 tbsps butter (1/2 stick)	4 celery stalks, diced
1 large onion, diced	1 ¼ teaspoons salt
6 cups whole-wheat bread cubes	1 ¼ cups water

6 cups white bread cubes (about 12 slices)

In a 5 to 8 quart Dutch oven over medium heat, in hot butter or margarine, cook celery and onion until tender stirring occasionally. Remove Dutch oven from heat; add bread cubes, water, Medjools and salt; toss to mix well. Use to stuff 12 to 16 pound turkey. Or spoon stuffing into 13x9" baking dish; cover with foil and bake in preheated oven at 325° 1 hour…or…until heated through. Yield: about 12 cups.

Medjool scones

2 cups all-purpose flour   3 tablespoons sugar
1 tablespoon baking powder   1 teaspoon salt
¾ cup pitted Medjools, chopped   2 eggs
6 tablespoons butter or margarine (3/4 stick)
2/3 cup milk

Preheat oven to 425°. Mix flour, sugar, baking powder and salt together. Chip butter into mix until mixture resembles coarse crumbs. In small bowl, beat eggs. Reserve 1 tablespoon beaten egg for brushing on batter later. Stir Medjools and milk into remaining beaten eggs. Stir egg mixture into flour mixture just until ingredients are blended. Grease 8"x8" baking pan.

Spoon Medjool mixture into prepared pan; brush top with reserved beaten egg. Bake 20 minutes or until golden. Cool in pan on wire rack 10 minutes. Remove from pan; cut in half then cut each half crosswise into thirds to make 6 rectangles. Cut each rectangle diagonally in half to make 12 rectangles. Serve warm. Just before serving, reheat if desired.

Medjool and nut chiffon pie

1 envelope unflavored gelatin   ½ cup sugar
½ teaspoon salt   4 egg yolks
¼ cup chopped Medjool dates   ¼ cup chopped walnuts
1 teaspoon grated lemon peel   4 egg whites
½ cup whipped cream, whipped   ½ cup sugar (2nd)
          1 9-inch baked, cooled, pastry shell

In saucepan, combine gelatin, ½ cup sugar and salt. Beat egg yolks, Medjools and walnuts and add 2/3 cup water. Stir into gelatin mixture. Cook and stir over medium heat just until mixture comes to boiling and gelatin is dissolved. Remove from heat and stir in peel. Chill, stirring occasionally until mixture is partially set. Beat egg whites until soft peaks form. Gradually add ½ cup sugar beating until stiff peaks form. Fold in gelatin mixture. Fold in whipped cream. Pile into cooled pastry shell. Chill until firm.

Medjool muffins II

¾ cup chopped Medjool dates   1 ¾ cups sifted all-purpose flour
¼ cup sugar   2 ½ teaspoons baking powder
¾ teaspoon salt   1 well-beaten egg
¾ cup milk   1/3 cup melted shortening

Sift dry ingredients including Medjools into bowl; Combine egg, milk and shortening. Add all to dry ingredients. Stir until dry ingredients are moistened. Fill muffin pans 2/3 full. Bake at 400° for 20 to 25 minutes.

Stuffed Medjool drops

35 pitted Medjool dates   35 walnut quarters
6 tablespoons butter, softened   1 egg
1/3 cup packed light brown sugar ¾ cup flour
¼ teaspoon baking powder   ¼ teaspoon soda
¼ teaspoon nutmeg   ¼ cup sour cream
1 ¼ cup confectioner's sugar   ¼ tsp vanilla extract

Pit Medjools and stuff with walnuts. Cream 2 tablespoons softened butter and brown sugar in bowl until light. Beat in egg. Add mixture of flour, baking powder, soda and nutmeg alternately with sour cream. Fold in stuffed dates. Drop coated Medjools 1 at a time onto greased cookie sheet. Bake at 400° for 8 to 10 minutes. Cool on wire rack. Heat remaining ¼ cup butter in saucepan until golden, stirring constantly. Remove from heat. Beat in confectioner's sugar, vanilla and enough water to make of spreading consistency. Frost cooled cookies.

Medjool pinwheels

2 ½ cups chopped Medjool dates   2 cups sugar
1 cup chopped nuts   1 cup shortening
1 cup packed brown sugar   3 eggs, beaten
4 cups flour   1 teaspoon soda
¾ teaspoon salt

Simmer Medjools with 1 cup water and 1 cup sugar in saucepan until tender. Stir in nuts. Cool. Cream shortening, brown sugar and remaining 1 cup sugar in bowl until fluffy. Add eggs; mix well. Add mixture of flour, soda and salt; mix well. Divide into 4 portions. Roll each into 9x11-inch rectangle on floured surface. Spread with cooled Medjool mixture. Roll as for jelly roll. Cut into 1-inch slices. Place on ungreased cookie sheet. Bake at 350° for 10 to 12 minutes or until golden brown. Cool on wire rack.   Yield: 12 dozen.

Section IV
Medjool Date Growers

Bard Date Company
5635 East Gila Ridge Road
Yuma Arizona, 85365-7630
928-344-2825
Fax: 928-341-9911
Email: info@barddate.com

Brown Date Gardens
69-245 Polk Street
Thermal, CA 92274
760-397-4309
www.browndategarden.com
Email: ted@browndategarden.com

California Desert Dates
90-785 Avenue 81
Thermal, CA 92274
760-397-2811
Fax: 760-397-1517

California Redi-Date LLC
P.O. Box 728
87-500 Airport Blvd.
Thermal, CA 92274-0728
760-399-5026
www.desertvalleydate.com

Cal Sungold, Inc.
82-291 Avenue 61
Thermal, CA 92274
760-399-5646
www.calsungold.com
Email: info@calsungold.com

Dateland Date Gardens
P.O. Box 3241
Interstate 8 & Milepost #67
Dateland, AZ 85333
928-454-2772
www.dateland.com
dateorderadmin@dateland.com
Email: datelandpalms@wildblue.net

Desert Valley Date, Inc.
86-740 Industrial Way
Coachella, CA 92236
760-398-0999
Email: sales@desertvalleydate.com

DatePac LLC
2375 E. 24th Street
Yuma, AZ 85364
928-314-0655
www.datepac.com

Dave's Medjools
1198 Perez Road
Winterhaven (Bard) CA 92283
760-572-2262

Dulin Date Gardens
5635 E. Gila Ridge Road
Yuma, AZ 85365
928-344-2685
Email: info@barddate.com

Fred Jaime Medjools
1330-A Perez Road
Winterhaven, CA 92283
760-572-2093

Gaymar Medjool Date Gardens
1178 Flood Road
Bard, CA 92222
760-572-0439

Hadley Date Gardens
83-555 Airport Blvd.
Thermal, CA 92274
760-399-5191
Email: sdougherty@hadleys.com

Imperial Date Gardens, Inc.
1517 York Road
P.O. Box 100
Bard, CA 92222
1-800-301-9349 / 760-572-0277
www.imperialdate.com
Email: impdategarden@aol.com

Jaime Date Farm
1449 Parkman Road
Winterhaven, CA 92283
619-572-0118

Jewel Date Company
48-440 Prairie Drive
Palm Desert, CA 92260
760-399-4474
Email: jeweldate@aol.com

Leja Farms
52-500 Van Buren Street
Coachella, CA 92236
760-398-8702
Email: lejafarms@aol.com

Martha's Gardens LLC
9747 9¾ Ave. E
P.O. Box 1034
Yuma, AZ 85366
928-726-8831
Email: contact@marthasgardensdatefarm.com

Oasis Date Gardens
59-111 Highway 111
P.O. Box 757
Thermal, CA 92274
760-399-5665
1-800-827-8017
www.oasisdategardens.com
Email: oasisdategardens@oasisdate.com

Pato's Dream Date Gardens
60-499 Highway 86
Thermal, CA 92274
760-399-5669
Email: patozdrm@aol.com

Pyramid Date Gardens
50-503 Van Buren
Coachella, CA 92236-9750
760-398-5182
www.datesaregreat.com

Ranchero DeLux Organic Date Gardens
Rt 1, 1452 Parkman Road
Winterhaven, CA 92283
760-572-0168
Email: medjools@beamspeed.net

Royal Medjool Date Gardens
833 E. Plaza Circle
Suite 200
Yuma, AZ 85364
928-726-0901

Seaview Packing, Inc.
86-235 Avenue 52
Coachella, CA 92236
760-398-8850
Fax: 760-398-8851
www.seaviewsales.com
Email: dennis@seaviewsales.com

Sun Date LLC
82-215 Avenue 50
Coachella, CA 92236
760-398-6123
Email: sundate@hotmail.com

Sun Garden Growers
1455 Hagberg Road
P.O. Box 190
Bard, CA 92222
760-572-0088
Email: sggrowers@aol.com

Western Date Ranches
1455 Hagberg Road
P.O. Box 190
Bard, California 92222
760-572-0088

Winterhaven Ranch
15621 Computer Lane
Huntington Beach, CA 92649
714-892-5586

Medjool Date Retailers and Gift-Pack Vendors

Basket Creations & More, LLC
245 South Main Street
Yuma, AZ 85364
928-341-9966
info@barddate.com

Berryman Farms of Bard
1197 Ross Road, Route 1
Winterhaven, CA 92283
760-572-0260

Brown Date Gardens
69-245 Polk Street
Thermal, CA 92274
760-397-4309
www.browndategarden.com
Email: ted@browndategarden.com

California Fruit Depot
10850 Redbank Road
Bakersfield, CA 93307
Office: 661-366-6303
ben@calfruitdepot.com

California Redi-Date LLC
P.O. Box 728
87-500 Airport Blvd.
Thermal, CA 92274-0728
760-399-5026
www.desertvalleydate.com

Cal Sungold, Inc.
82-291 Avenue 61
Thermal, CA 92274
760-399-5646
info@calsungold.com

Dateland Date Gardens
P.O. Box 3241
Interstate 8 & Milepost #67
Dateland, AZ 85333
928-454-2772
www.dateland.com
dateorderadmin@dateland.com
Email: datelandpalms@wildblue.net

Datilera del Desierto
Business Offices:
Datilera del Desierto
Zaragoza Nm. 1938
Altos-2 Mexicali, Baja California
Address USA:
P.O. BOX 1629, Calexico, CA
92232 U.S.A.
Ph. 52 65 52 4541 - (Mexico)
Fax. 52 65 52 6363 - (Mexico)
jluken@datiluken.com

Dave's Medjools
1198 Perez Road
Winterhaven, CA 92283
760-572-2262

Desert Valley Date
86-740 Industrial Way
Coachella, CA 92236
760-398-0999
sales@desertvalleydate.com

Fred Jaime Medjools
1330-A Perez Road
Winterhaven, CA 92283
760-572-2093

Gaymar Medjool Date Gardens
1178 Flood Road
Bard, CA 92222
760-572-0439

Hadley Fruit Orchards
48980 Seminole Drive
Cabazon, CA 92230
Cabazon exit off Interstate 10
800-854-5655
www.hadleyfruitorchards.com

Imperial Date Gardens, Inc.
1517 York Road
Bard, CA 92222
760-572-0277
www.imperialdate.com
Email: *imperialdate@digitaldune.net*

Jaime Date Farm
1449 Parkman Road
Winterhaven, CA 92283
619-572-0118

Leja Farms
52-500 Van Buren Street
Coachella, CA 92236
760-398-8702
Email: *lejafarms@aol.com*

Martha's Gardens L.L.C.
9747 9¾ Ave. E
P.O. Box 1034
Yuma, AZ 85366
928-726-8831
Email: *contact@marthasgardensdatefarm.com*

Oasis Date Gardens
59-111 Highway 111
P.O. Box 757
Thermal, CA 92274
760-399-5665
1-800-827-8017 Mail Order
www.oasisdategardens.com
Email: *oasisdategardens@oasisdate.com*

Pato's Dream Date Gardens
60-499 Highway 86
Thermal, CA 92274
760-399-5669
Email: *patozdrm@aol.com*

Ranchero DeLux Organic Date Gardens
Rt 1, 1452 Parkman Road
Winterhaven, CA 92283
760-572-0168
Email: *medjools@beamspeed.net*

Royal Medjool Date Gardens
P.O. Box 930
Bard, California 92222
928-726-0901
Fax: 928-726-9413
Email: *rmdates@worldnet.att.net*

Shields Date Gardens
80-225 Highway 111
Indio, CA 92201-6599
800-414-2555 Mail Order
760-347-0996 Mail Order
760-347-7768 Store
FAX: 760-342-3288
Email: *shieldate@aol.com*
www.shieldsdategarden.com

Sun Date LLC
82-215 Avenue 50
Coachella, CA 92236
760-398-6123
Email: *sundate@hotmail.com*

Sun Garden Growers
1455 Hagberg Road
P.O. Box 190
Bard, CA 92222
760-572-0088
Email: *sggrowers@aol.com*

The Peanut Patch
4322 E. County 13th St.
Yuma, AZ 85365
928-726-6292
www.thepeanutpatch.com